essentials

Help Your 5–7 Year Old Tell the Time

See back cover for address.

Help Your 5–7 Year Old Tell the Time

Ken Adams

PARENTS' ESSENTIALS

Published in 2001 by
How To Books Ltd, 3 Newtec Place,
Magdalen Road, Oxford OX4 1RE, United Kingdom
Tel: (01865) 793806   Fax: (01865) 248780
email: info@howtobooks.co.uk
www.howtobooks.co.uk

British Library Cataloguing in Publication Data.
A catalogue record for this book is available from the British Library.

Cover design by Shireen Nathoo Design
Produced for How To Books by Deer Park Productions
Typeset by PDQ Typesetting, Newcastle-under-Lyme, Staffordshire
Printed and bound in Great Britain by The Baskerville Press Ltd.

NOTE: The material contained in this book is set out in good faith for general guidance and no liability can be accepted for loss or expense incurred as a result of relying in particular circumstances on statements made in the book. Laws and regulations are complex and liable to change, and readers should check the current position with the relevant authorities before making personal arrangements.

ESSENTIALS *is an imprint of*
How To Books

Contents

Preface

Helping your child at home has become more and more the accepted norm, and there has been a large increase in the number of work books to be used as back-up to schoolwork. Given two children of similar abilities, the one who has help and encouragement at home is likely to do better at school. However, this will depend on how well a parent understands what they need to do. This requires a book to spell out **how** to help your child in key areas, as well as simply providing practice material.

One of the key areas is 'Telling the Time'. Clearly, this is of considerable practical use, but also it has many ramifications through maths number work. What might seem to be a simple learning of telling the time involves several maths skills and some problem-solving ability. This book aims to advise parents and teachers on the best way to progress, so that understanding and memorisation is relatively easily achieved. There is much practice for a young child that is staged in step-wise progression as well as vital advice throughout.

1 How Your Child Learns

By the age of five years a child has a fairly wide bank of knowledge and understanding in memory. This is mainly built up of real-life understanding – linking different aspects of objects within memory, the recognition of 'a bus' as an object and what its use is, for example. The deeper the understanding and the wider the knowledge, the more links there are to be made when learning in the more abstract environment of school. Your child builds on what he has already learnt, so the more 'pegs' there are to hang new information on the better, and learning will be easier. 'Finding out' for a young child is a crucial exercise. Academic works often depends on the success pre-school of such discovery about their world.

LINKING NEW LEARNING IN STEPS

Another important principle in learning is that what is learnt is close enough to what a child knows for understanding to be achieved. At this age a child is used to dealing with and solving simple real-life problems like 'Here is a clock – what does it tell me? What is it for?'. Telling the time, however, deals with abstract concepts like counting to at least 30, half, quarter, minutes past, minutes to, so learning must be linked carefully to what a child knows or it will fail.

Learning must also occur in carefully graded steps, so that your child finds it easy to move from one step to the next, learning both quickly and easily.

MOVING FROM THE SIMPLE TO THE MORE COMPLEX

In all learning it is also important that you start from a simple base and move on in steps to more complex ideas. Telling the time, for example, is a combination of relatively complex ideas for a 5–7 year old. Apart from being able to count well, which is essential, your child must be able to recognise the 'o'clock', the 'half' point (half-past) and the two 'quarter' points: quarter past and quarter to. Finally, they must be able to work out minutes to and minutes past, and then write this down in an acceptable form, e.g., 6.45 a.m. or 6.45 p.m.

There are clear stages of complexity, or difficulty, in learning to tell the time, but if these are recognised and the learning is ordered in a common sense manner, your child will learn quickly and easily, whatever their ability.

THE ABILITY TO LEARN

Any parent or teacher would agree that children vary enormously in their ability to learn. John, who was at university entrance level in maths at age nine, learnt to tell the time at two years of age, to the minute: 'It is sixteen minutes to two,' he lisped to the astonished doctor at the baby clinic. Others are still not telling the time correctly at a much older age. One fourteen year old came to me and asked a little shamefully, 'Can you help me to learn how to tell the time? Nobody ever taught me.'

It depends of course on whether parents or teachers even bother to teach telling the time, whether or not they teach in an effective way, and not simply on ability to learn. This is the purpose of this book, to provide an effective plan for teaching telling the time, so that slow learners as well as quick learners can master the various skills that allow your child to both recognise the time on a clock and be able to use that knowledge in maths examples.

STAGES IN LEARNING

The following chapters move from exercises that ensure that your child can count, at least up to 60, and this counting is also related to the minutes marked out around the clock face. Counting in fives is very important (5 times table). Your child must then recognise the o'clocks, halves and quarters on the clock face. Practice given in this book on these areas is fairly extensive. Further chapters deal with minutes past and minutes to the hour, and how to write them down. Finally, telling the time enters the area of mathematics more through asking such questions as 'how long is it from 9.15 a.m. to 2.30 p.m.', and through simple problem solving.

2 Learning the O'clocks, Half-pasts and Quarters

RECOGNISING THE O'CLOCKS

The first aim is to make sure that your child can recognise the big hand or long hand and small or short hand. The o'clock used to be represented as a firm pattern in your child's mind: When the **big** hand is on the twelve it is the **o'clock**. You look at the **small** hand to tell you the **number o'clock**.

So,

long hand (big hand)

This is the 'o'clock.

A good idea is to cut out a cardboard clock and draw the big hand fixed to the o'clock position. Alternatively, you may decide to buy a cardboard or wooden clock face with two hands – a small hand and a big hand. The use of such simple aids can greatly speed the learning of telling the time. Pointing out the o'clock times and the two hands on clocks around the house can also help your child to understand the purpose of the clock face.

Also, a play-clock, if it is properly made, can demonstrate that the **small** hand moves only between the numbers when the **big** hand moves through the whole circle.

These early ideas are essential for your child to grasp before they move on to the next stage of learning. It is assumed that your child is able to copy the numbers from 1 to 12.

Practice at recognising the o'clocks

Write beneath each of these clocks what o'clock it is:

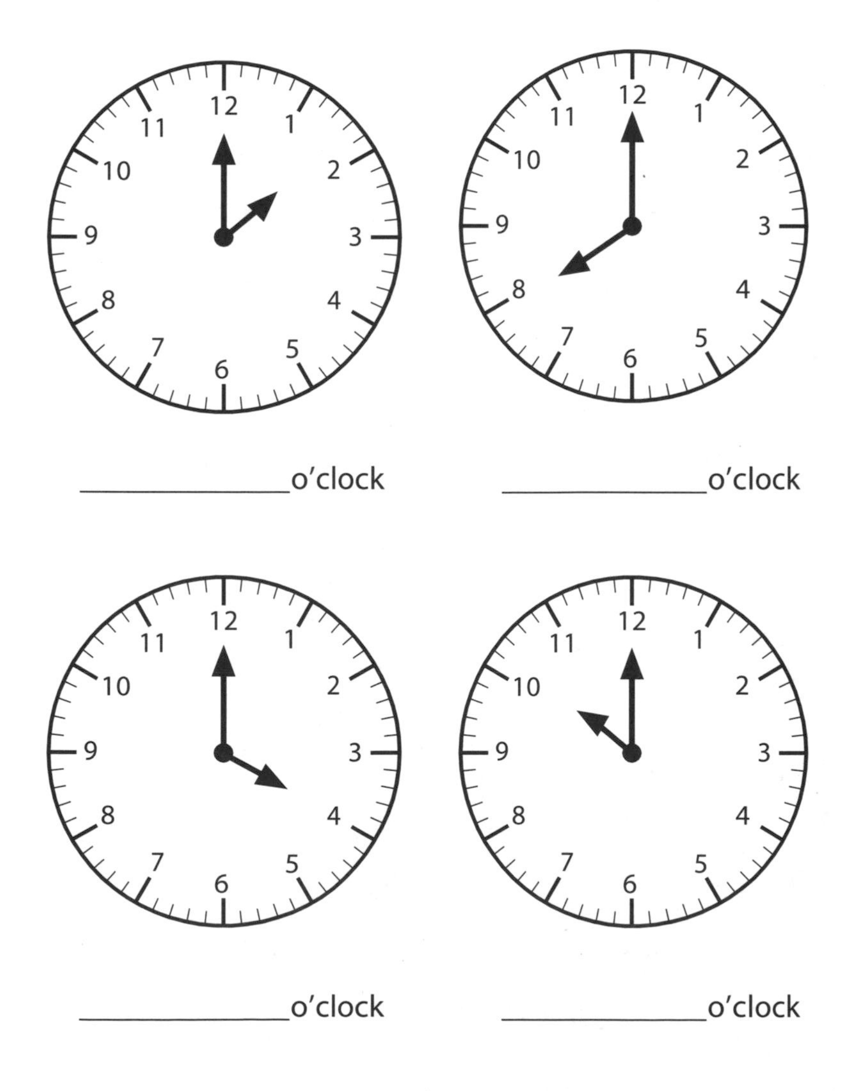

12
1
2
3
4
5
6
7
8
9
10
11
________o'clock
________o'clock
________o'clock
________o'clock
________o'clock
________o'clock

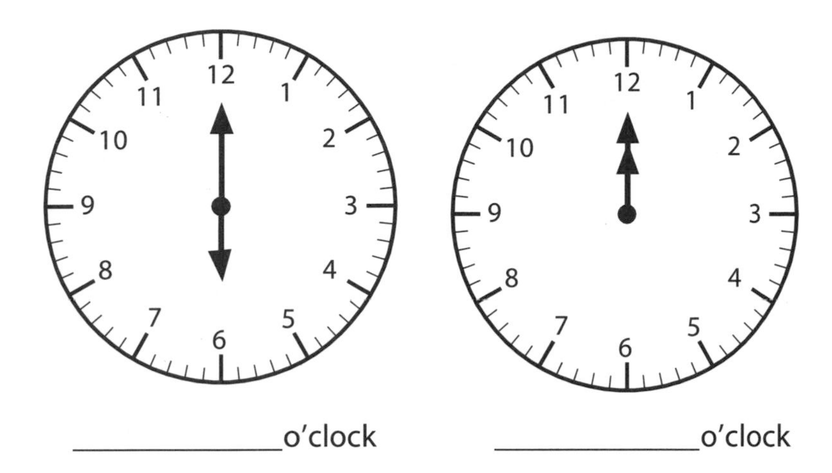

_______________o'clock _______________o'clock

RECOGNISING THE HALF-PASTS

As with the o'clocks, your child will need to recognise the pattern of the **long** hand being 'straight down' for half-past:

Next, they must recognise that the **small** hand will not now be pointing exactly at a number. The number before where the **small** hand points is the number to use:

Practice at recognising the half-pasts

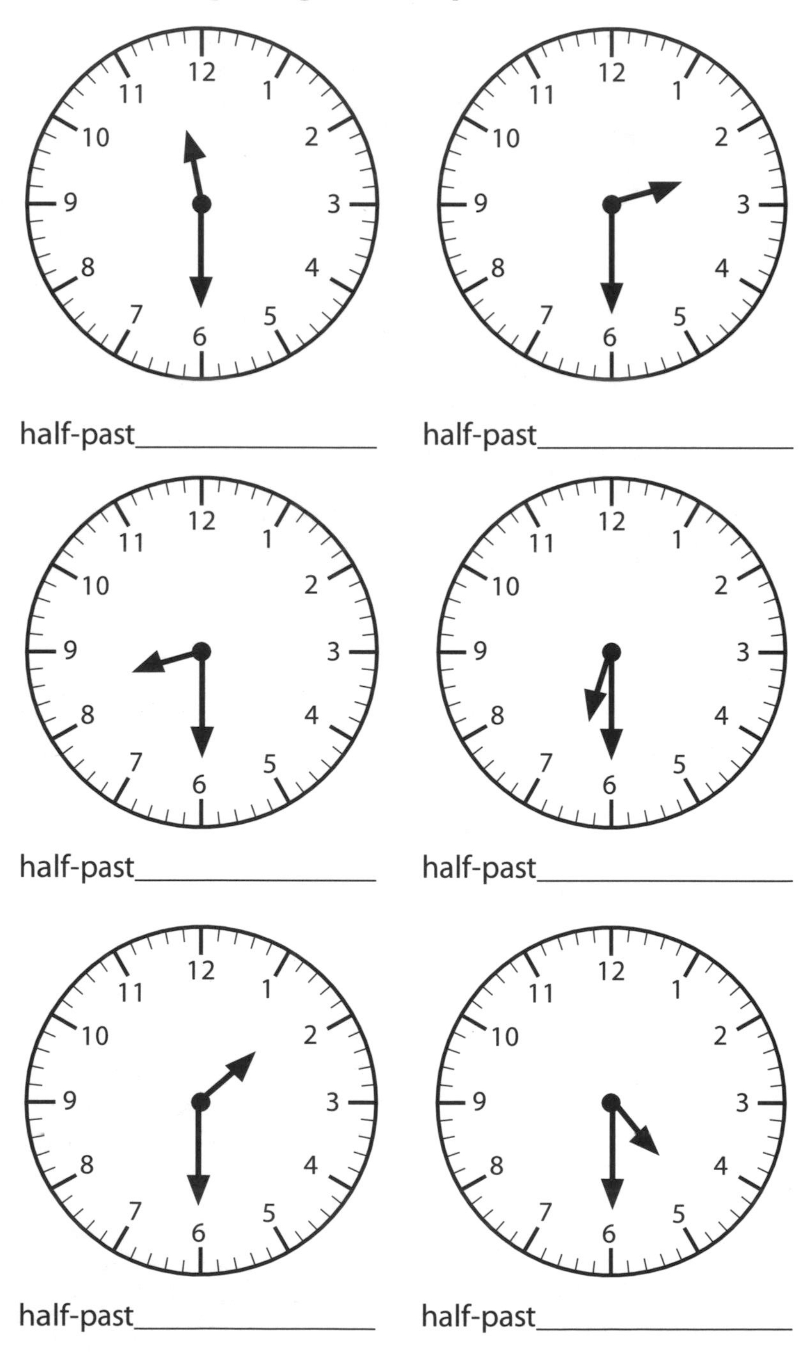

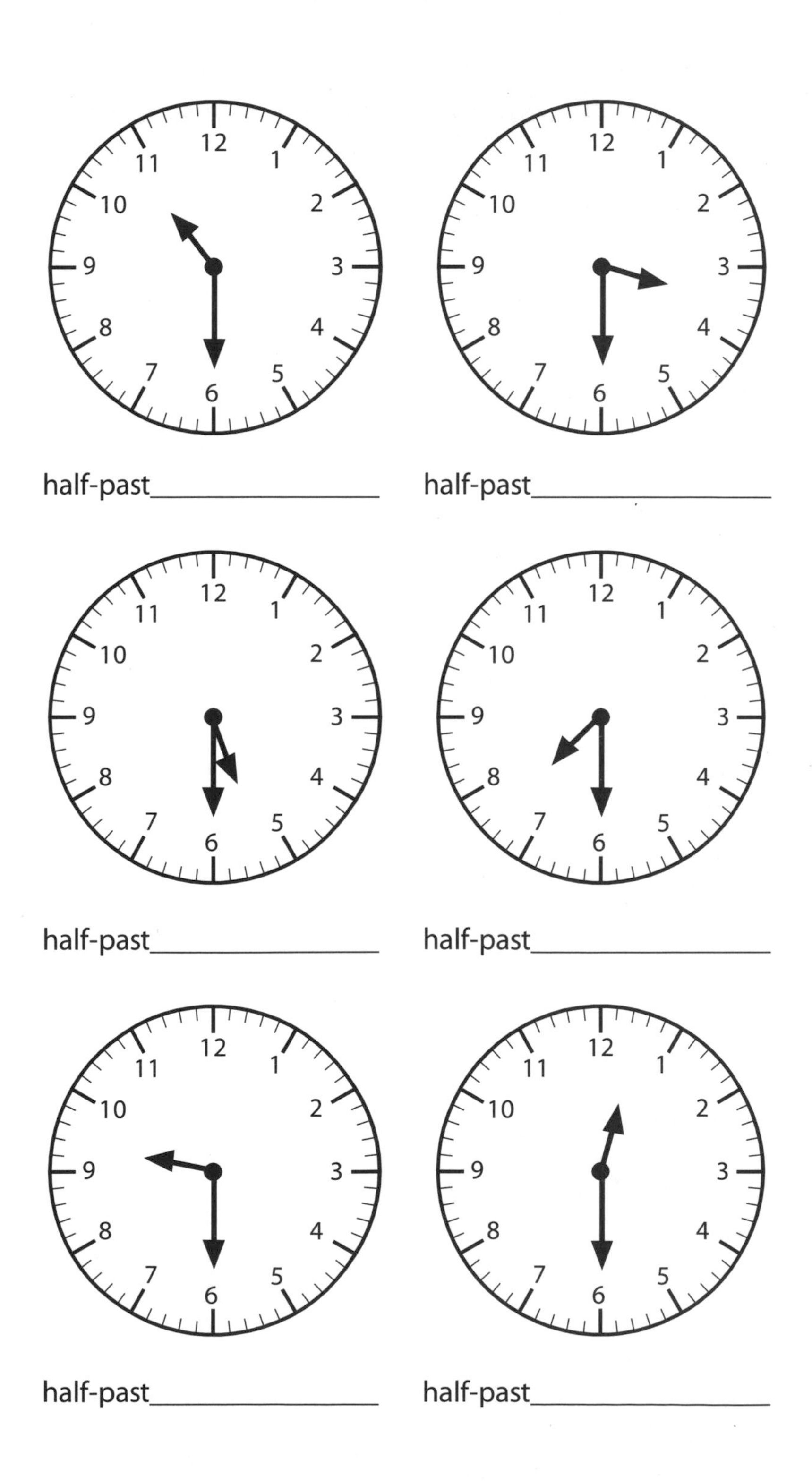
half-past
half-past
half-past
half-past
half-past
half-past

RECOGNISING THE QUARTER-PASTS AND QUARTER-TOS

Recognising the quarter-pasts should be taught separately from the quarter-to points, and then these should be practised together. This is a step-wise system that allows for better and faster learning.

Quarter-past

Your child needs to be aware of the quarter-past **pattern**, and also the **words** associated with this:

long hand

quarter-past________

There is a need at this stage to teach fractions. However, understanding the clock times does **help** with early fraction learning.

Practice at learning the quarter-pasts

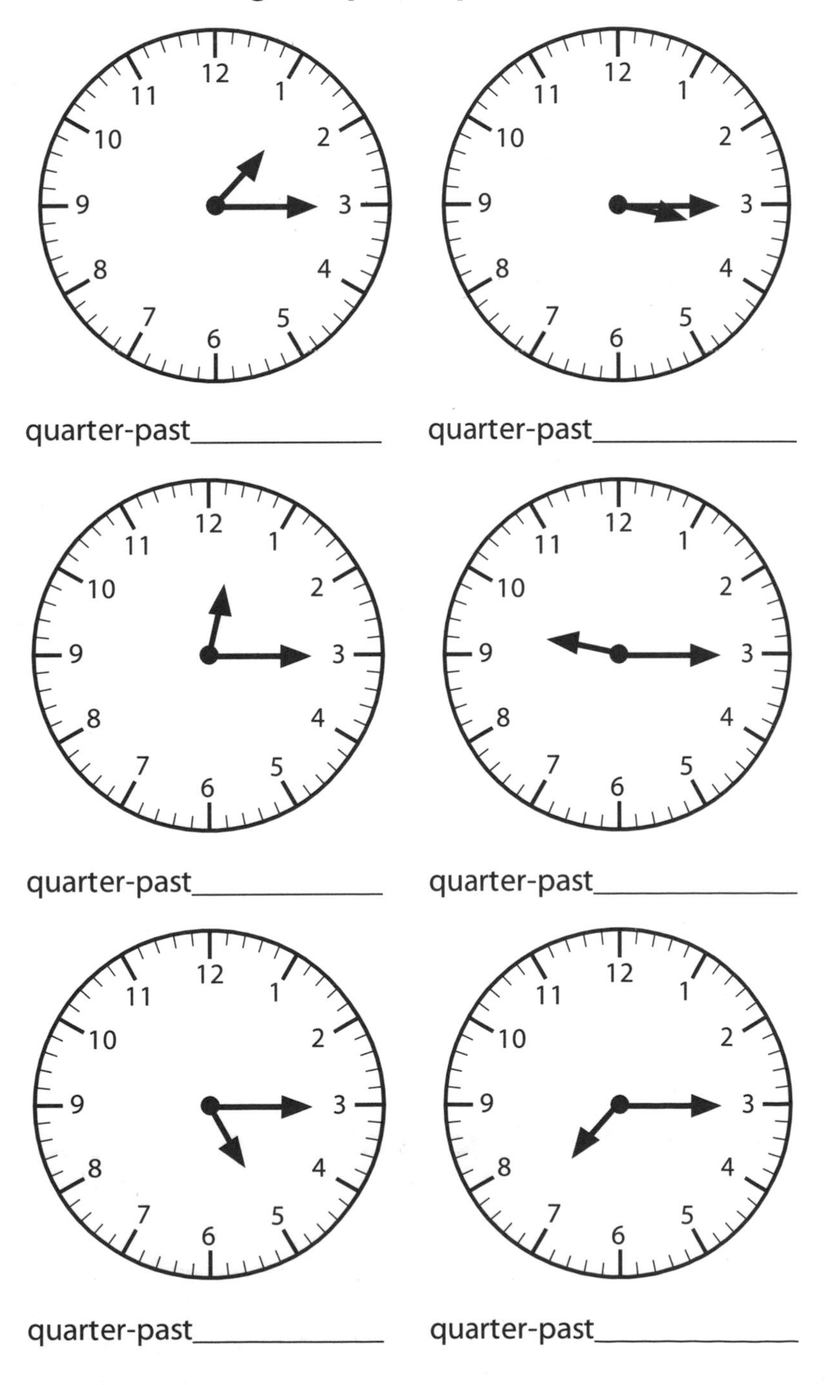

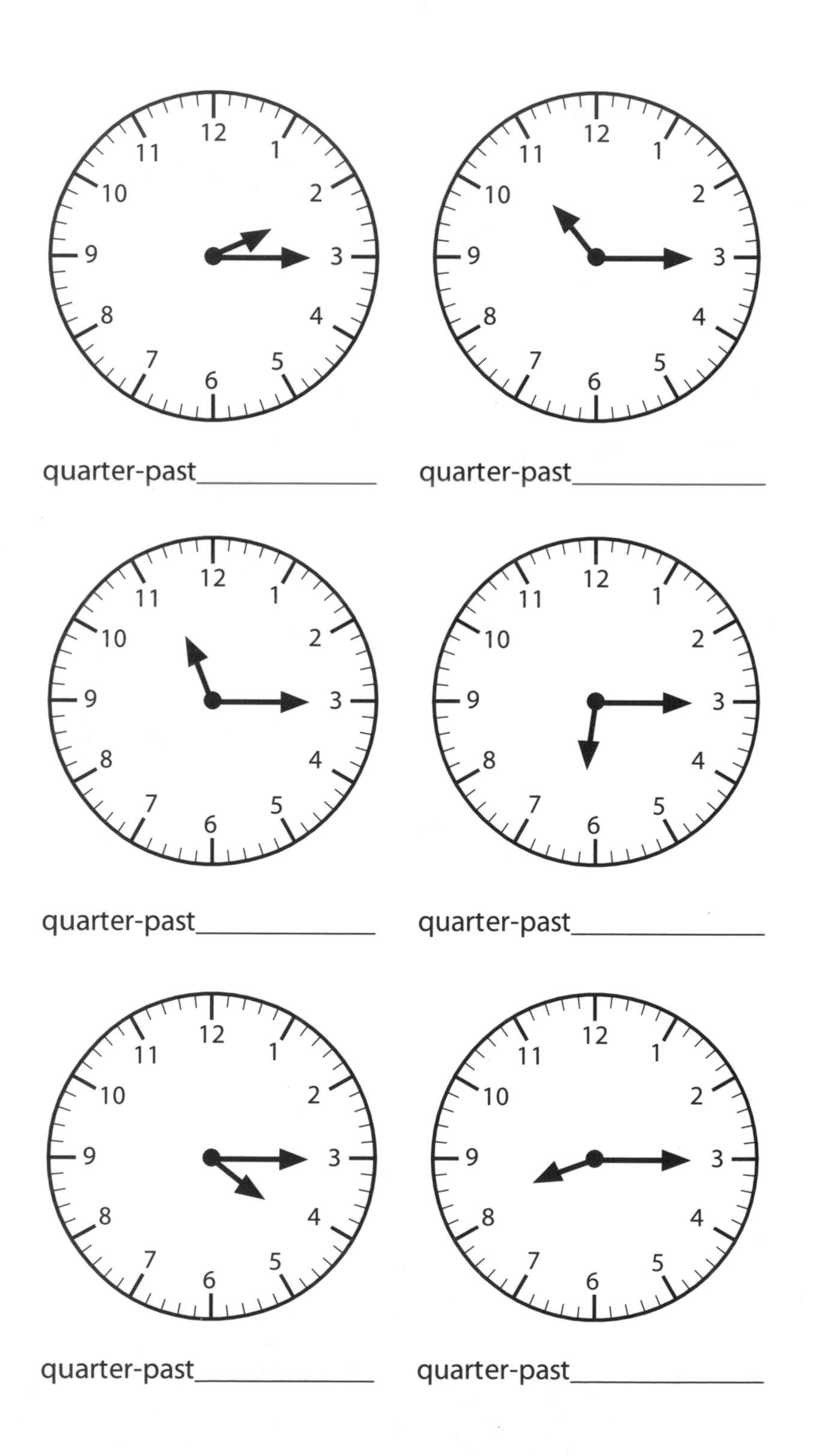
quarter-past______________
quarter-past______________
quarter-past______________
quarter-past______________
quarter-past______________
quarter-past______________

It needs to be emphasised that the number of the o'clock here is given by the number **before** the small hand:

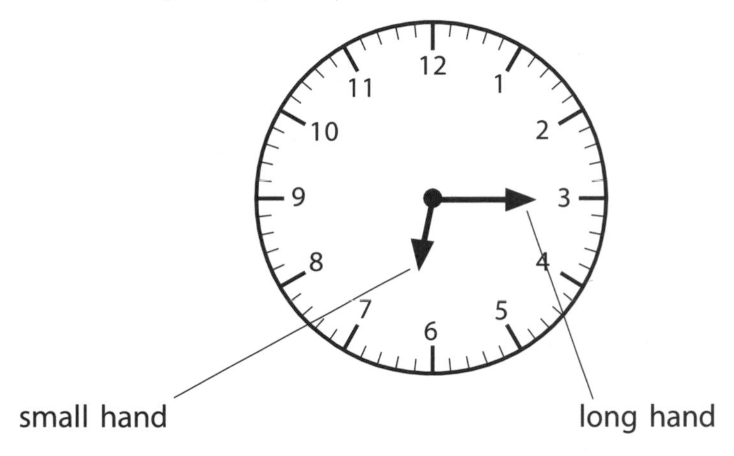

Practice at recognising the quarter-to times

The pattern with the long hand needs to be reinforced:

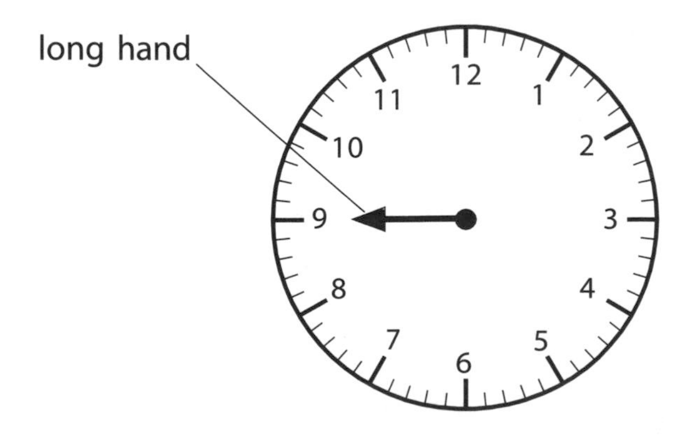

A further difficulty is that many small children become confused because they must now name the number **after** the position of the small hand, not the number before:

This anomaly over the hour hand needs to be emphasised as well as using practice to cement the idea:

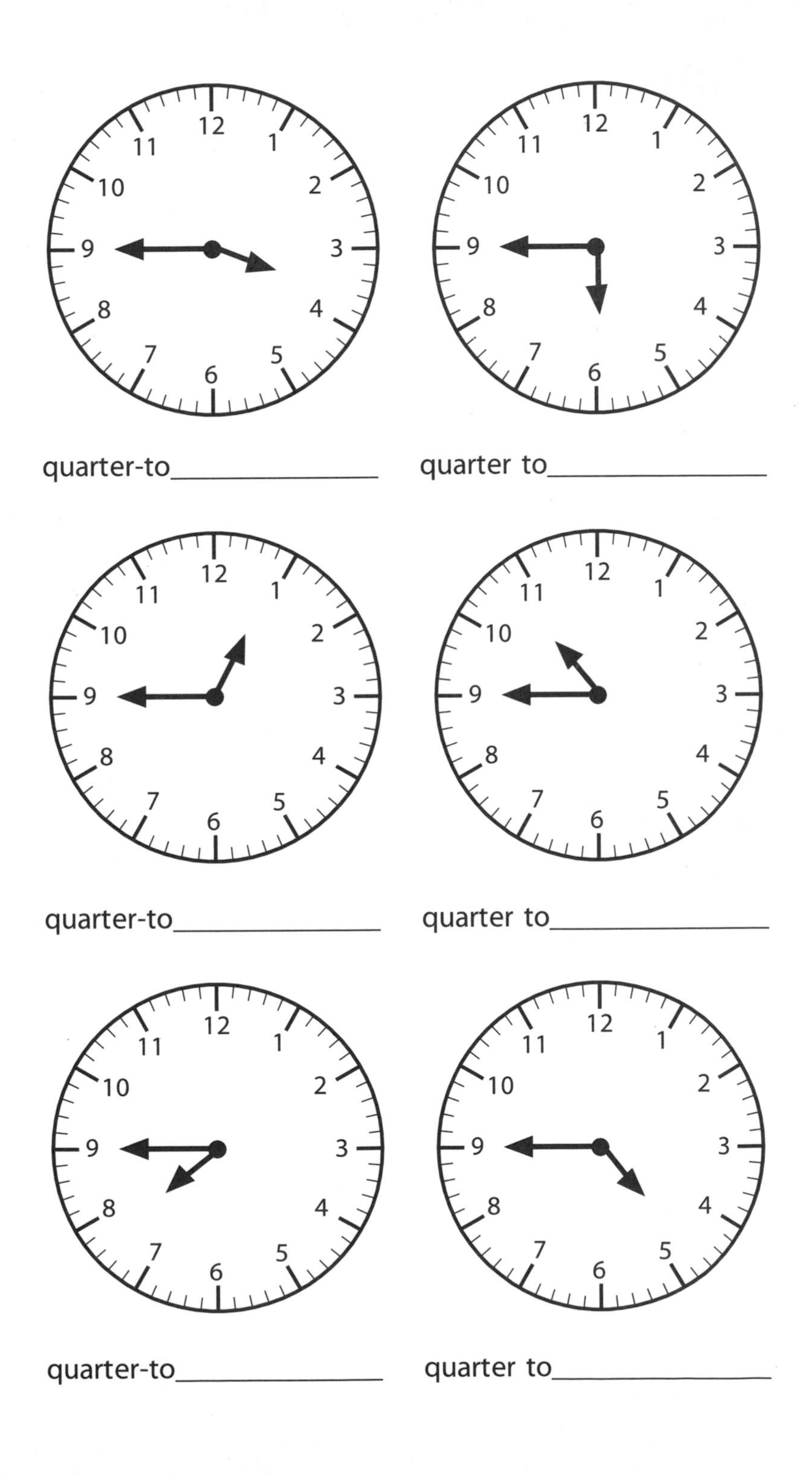

quarter-to________________ quarter to________________

quarter-to________________ quarter to________________

quarter-to________________ quarter to________________

3 Counting for Telling the Time

At this stage many children have a good knowledge of counting, but cannot relate easily to counting forwards and backwards around a clock face. In particular, many children find the meaning of the numbers, which represent the 5 times table, difficult to understand.

COUNTING TO 60

Basically, in telling the time, your child will need to be able to count to 60, for times like 7.46 to be written down and understood. Initially, though, counting to 30 is sufficient, but your child will need to be able to count up to 30 **back** from the 12, as well as forward (e.g., 26 minutes **to** 5, as well as 26 minutes **past** 5).

This means that there must be practice using a number line, forwards and backwards.

Practice on the number line up to 30

A (6), B (9), C (12), D (15), E (18), F (24), G (27)

0 3 6 9 12 15 18 21 24 27 30

How many is it to:

(a) to A ___________ (b) to B___________

(c) to C ___________ (d) to D___________

(e) to E __________ (f) to F__________

(g) to G __________

How many back from 30:

(a) to G __________ (b) to F__________ (c) to E__________

(d) to D __________ (e) to C__________ (f) to B__________

(g) to A __________

Practice in a circle to half-way, and back to half-way

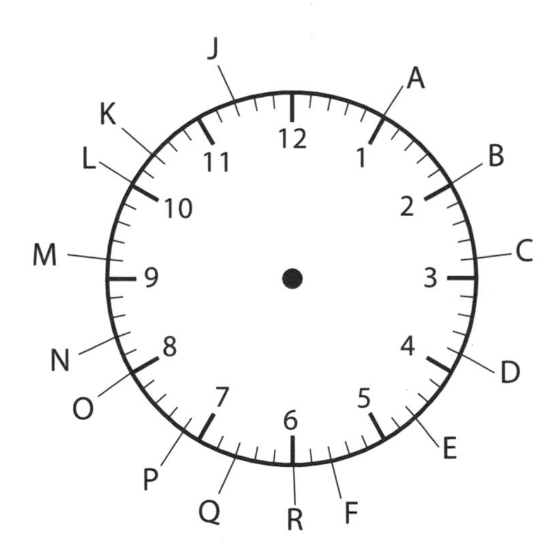

How far is it from the 12 point to:

(a) A __________ (b) B__________ (c) C__________

(d) D __________ (e) E__________ (f) F__________

How far is it from the 12 **back** to:

(a) J ____________ (b) K______________ (c) L______________

(d) M ____________ (e) N______________ (f) O______________

(g) P ____________ (h) Q______________ (i) R______________

COUNTING IN FIVES TO 30

Telling the time involves quite difficult counting, but also the recognition of 5 times patterns. For example, the 5 on the clock represents 25 minutes past, the 7 represents 35 minutes past or 25 minutes to. This confuses early learners considerably, so the counting elements and patterns must be thoroughly learned before there can be easy learning of the time.

Practice in counting in 5s along the number line

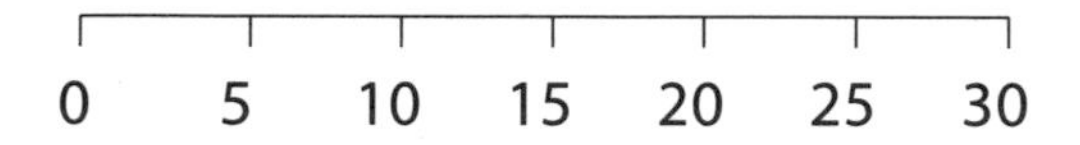

How many is:

(a) 3 fives___________ (b) 2 fives___________

(c) 4 fives___________ (d) 1 five___________

(e) 5 fives___________ (f) 6 fives___________

COUNTING ON THE CLOCK

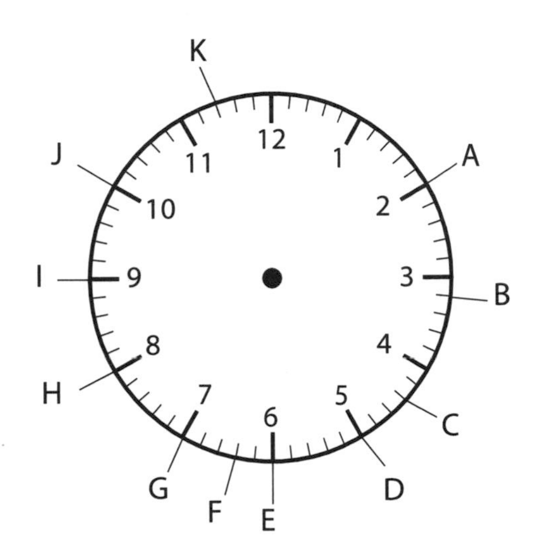

Count to:

(a) A __________ (b) B__________

(c) C __________ (d) D__________

(e) E __________ (f) F__________

(g) G __________ (h) H__________

(i) I __________ (j) J__________

COUNTING BACK ON THE CLOCK

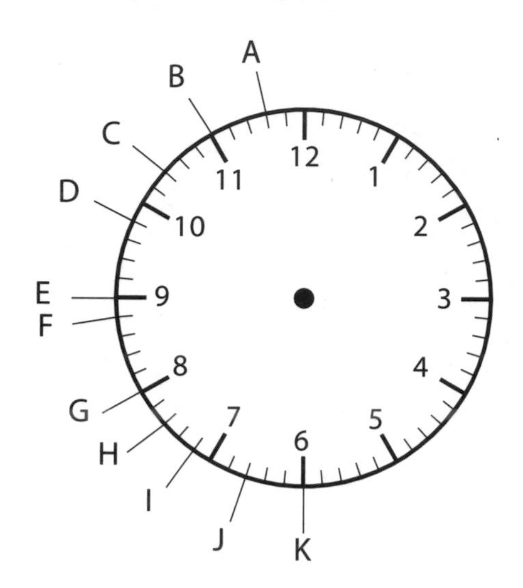

Count back from '12'

(a) to A ___________ (b) to B___________

(c) to C ___________ (d) to D___________

(e) to E ___________ (f) to F___________

(g) to G ___________ (h) to H___________

(i) to I ___________ (j) to J___________

(k) to K ___________

COUNTING TO 60 IN FIVES

This requires counting in fives through to 60; in other words, your child will need to know the 5 times table up to 12 x 5.

Practice in 5 times table

Work out these:

(a) 1 x 5 = ________ (b) 3 x 5 = ________

(c) 5 x 5 = ________ (d) 6 x 5 = ________

(e) 4 x 5 = ________ (f) 7 x 5 = ________

(g) 8 x 5 = ________ (h) 9 x 5 = ________

(i) 10 x 5 =________ (j) 11 x 5 = ________

(k) 2 x 5 = ________ (l) 12 x 5 = ________

How many?

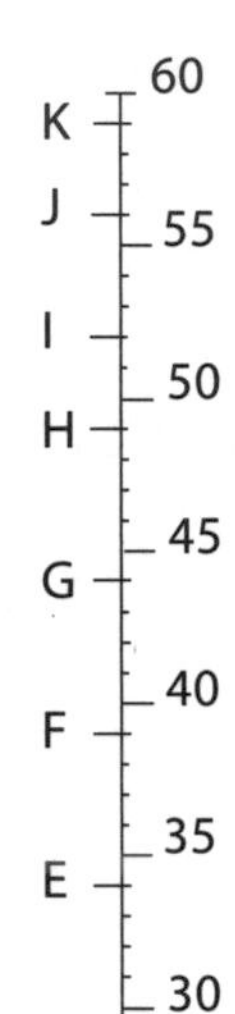

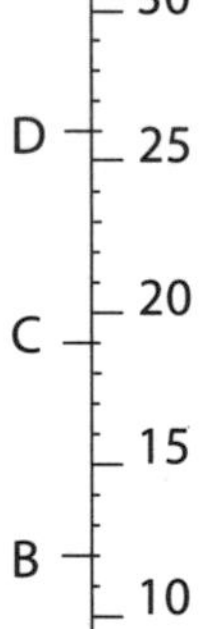

(a) to A __________ (b) to B__________

(c) to C __________ (d) to D__________

(e) to E __________ (f) to F__________

(g) to G __________ (h) to H__________

(i) to I __________ (j) to J__________

(k) to K __________

4 Learning to Write the Time

For a young learner, the many ways of representing the time can be very confusing. From simple to complex, the time can be represented as, for example:

Half past three

Quarter past three

Quarter to three

3.30 a.m. or 3.30 p.m.

3.15 a.m. or 3.15 p.m.

3.45 a.m. or 3.45 p.m.

and, in terms of the 24-hour clock:

03.30 or

15.30

Twenty-four hour times are dealt with in the next chapter. This chapter is concerned with helping your child to be able to write down any time in terms of a.m. and p.m.

RECOGNISING MINUTES TO AND PAST THE HOUR

Are these times **to** or **past** the hour?

Tick the correct one:

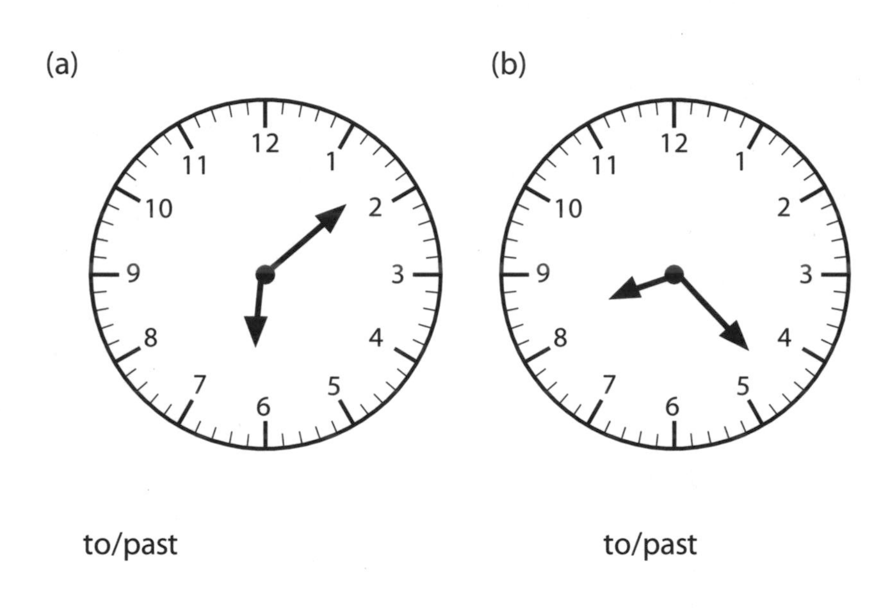

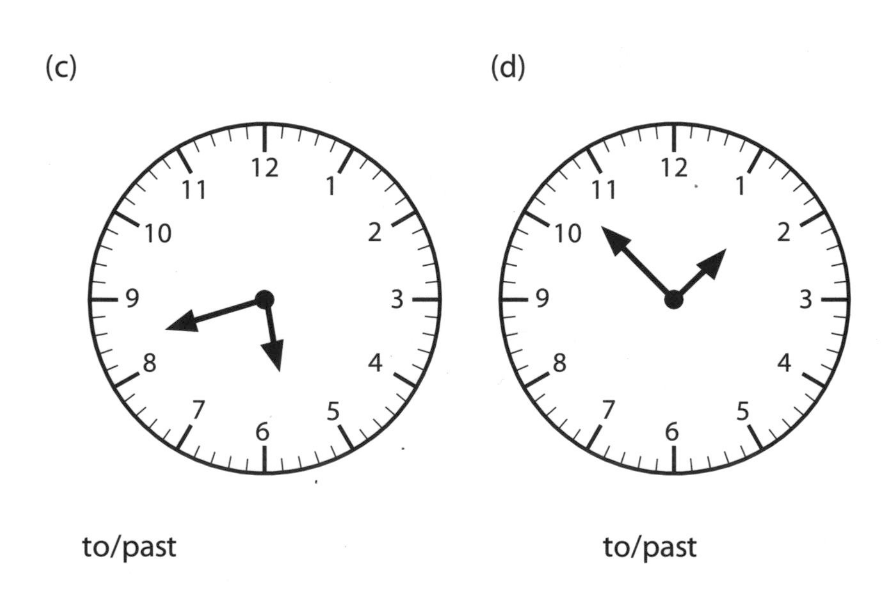

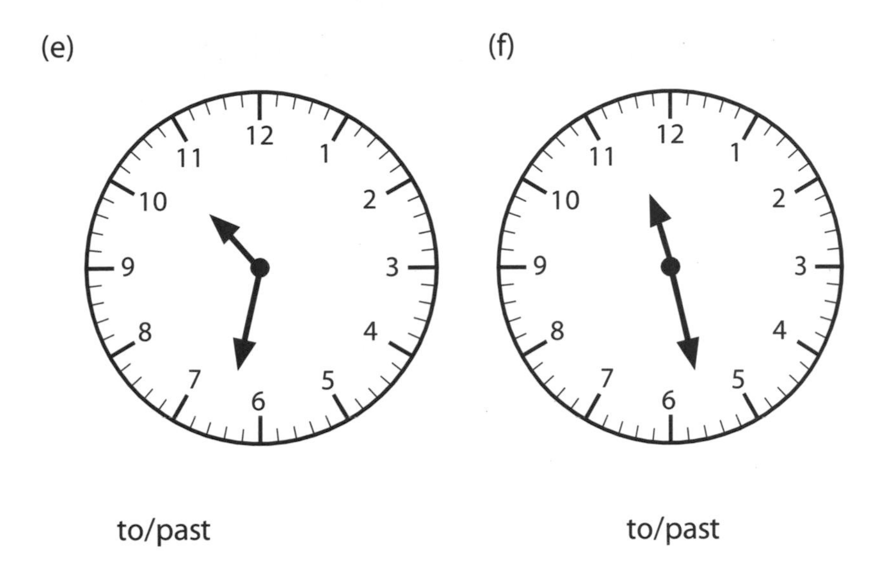

The next stage is to concentrate on the **past** the hour times:

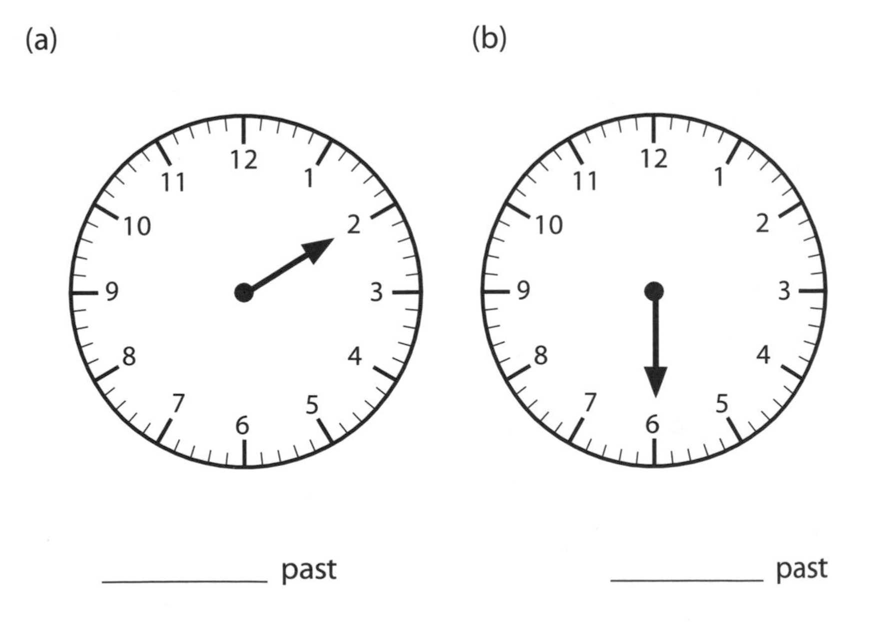

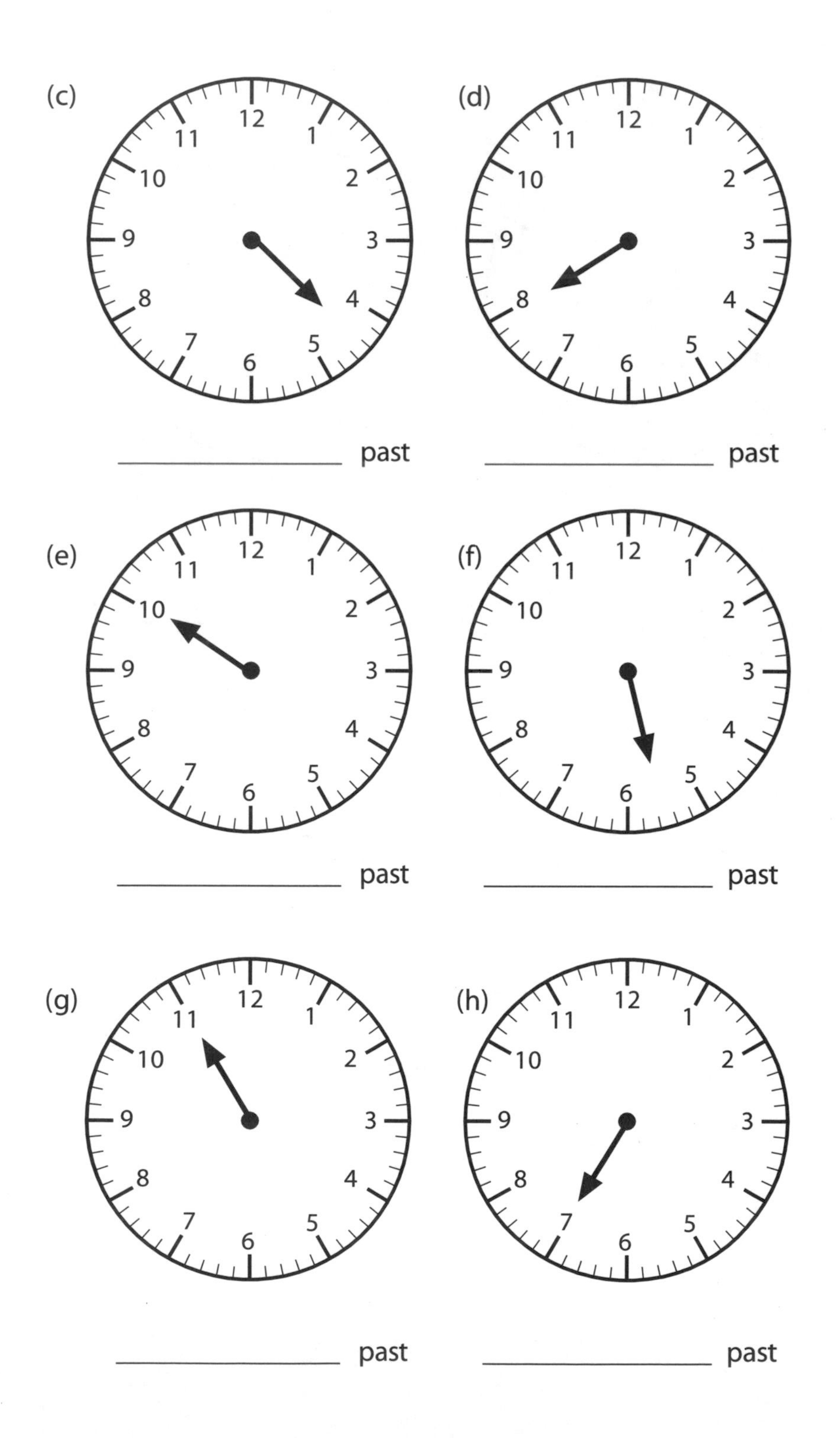
(c)
12
1
2
3
4
5
6
7
8
9
10
11
past
(d)
12
1
2
3
4
5
6
7
8
9
10
11
past
(e)
12
1
2
3
4
5
6
7
8
9
10
11
past
(f)
12
1
2
3
4
5
6
7
8
9
10
11
past
(g)
12
1
2
3
4
5
6
7
8
9
10
11
past
(h)
12
1
2
3
4
5
6
7
8
9
10
11
past

The second stage is to get your child to work on the minutes **to** the hour:

Write the time down as minutes to the hour

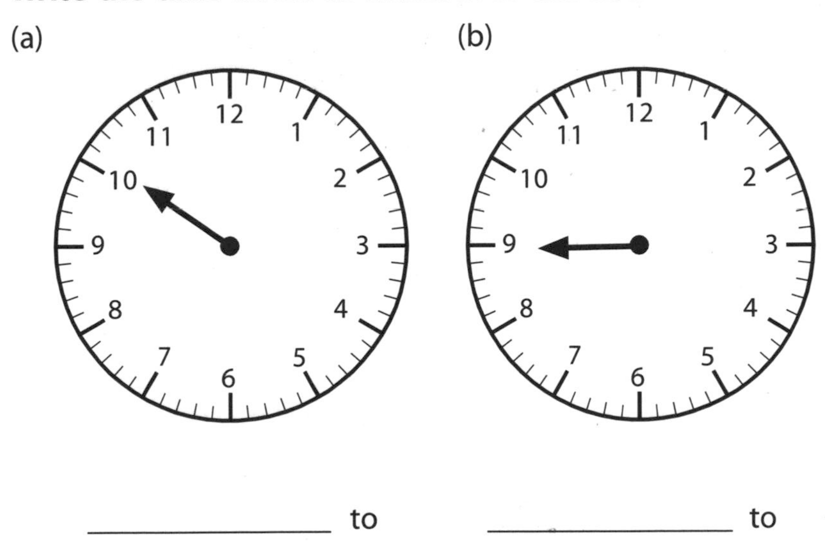

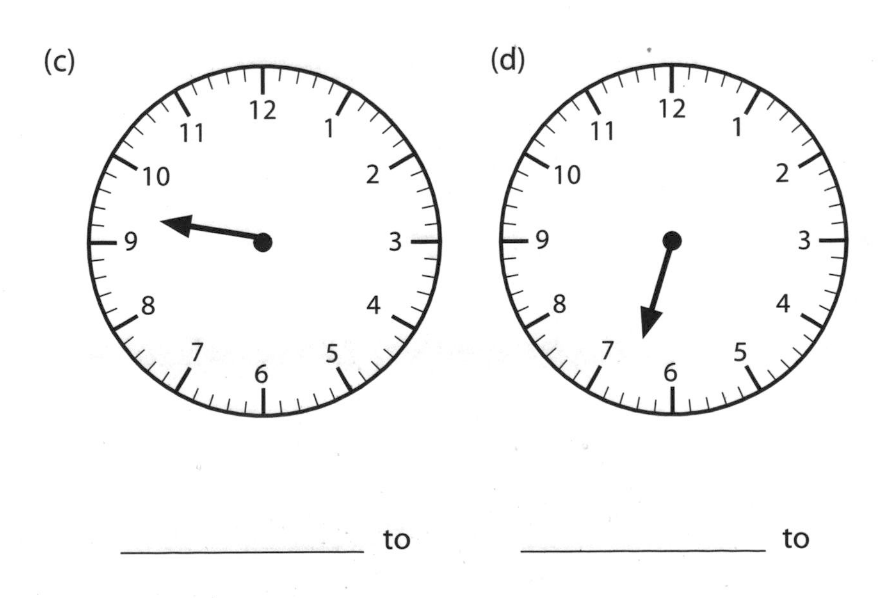

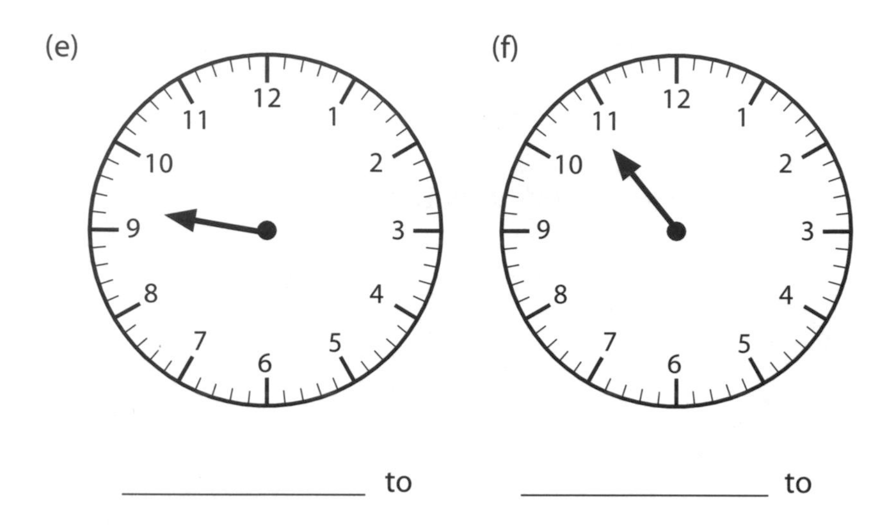

It is now necessary for your child to learn to write down these times in terms of a.m. and p.m. In combination, they should be able to count the minutes up to 30.

Write down these times. They are a.m. times (before 12 mid-day).

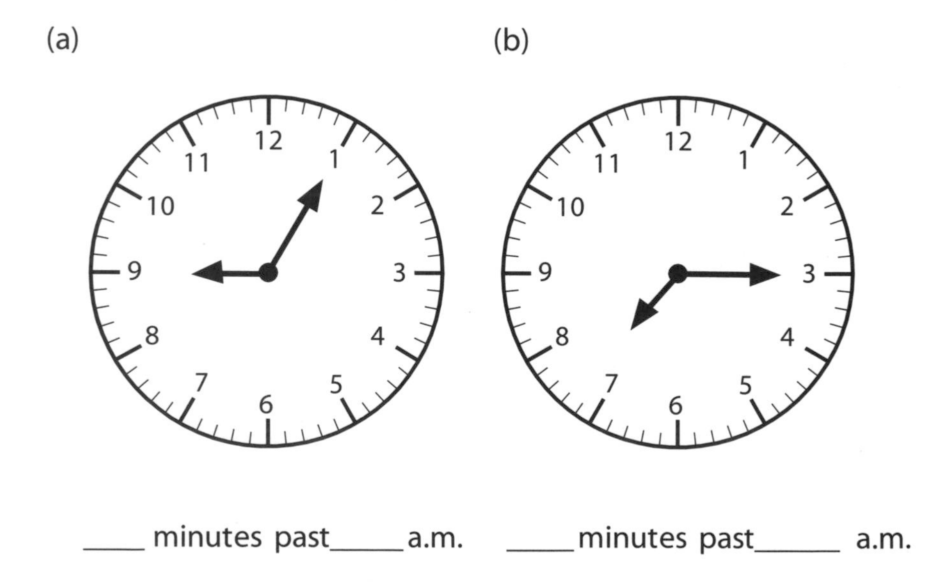

(c)

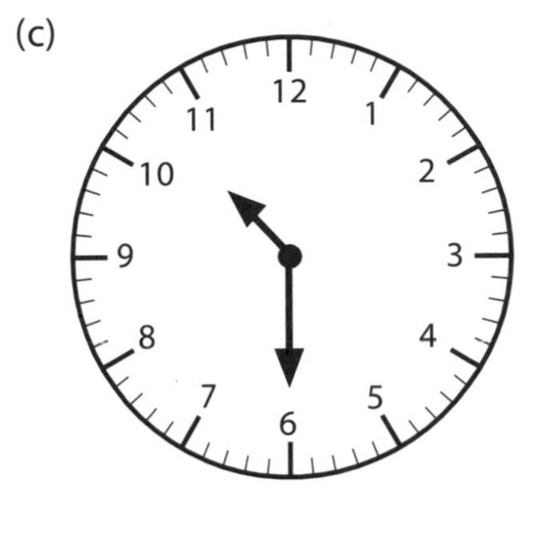

____ minutes past_____ a.m.

(d)

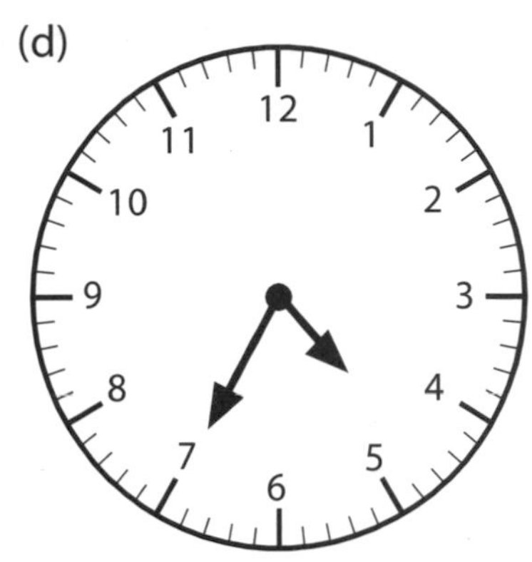

____minutes past_____ a.m.

(e)

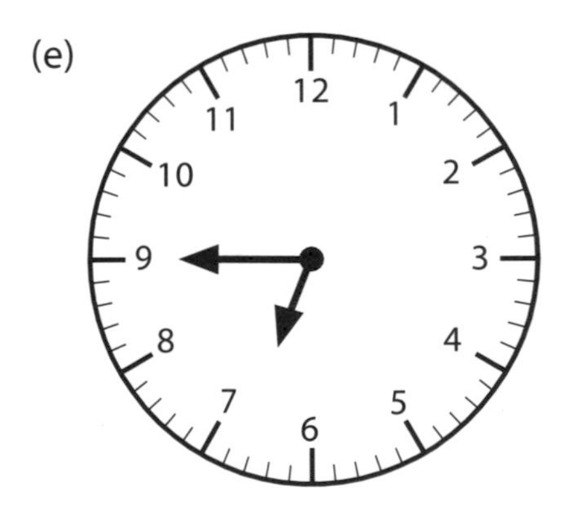

____ minutes past_____ a.m.

(f)

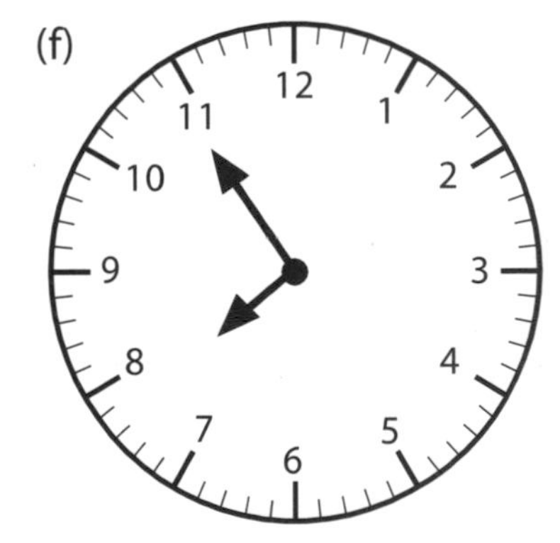

____minutes past_____ a.m.

Practice at writing down minutes to the hour

Write down these times:

(a)

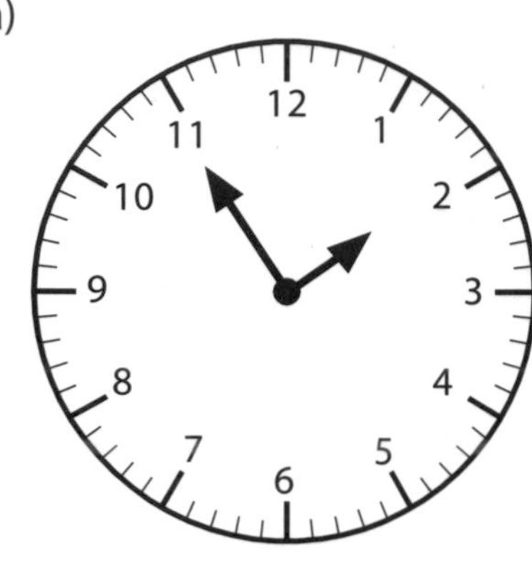

(b)

____ minutes to______ ____minutes to_______

(c)

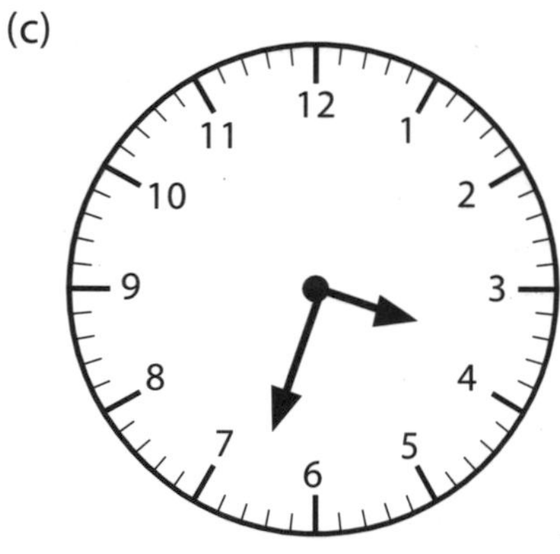

(d)

____ minutes to______ ____minutes to_______

(e) (f)

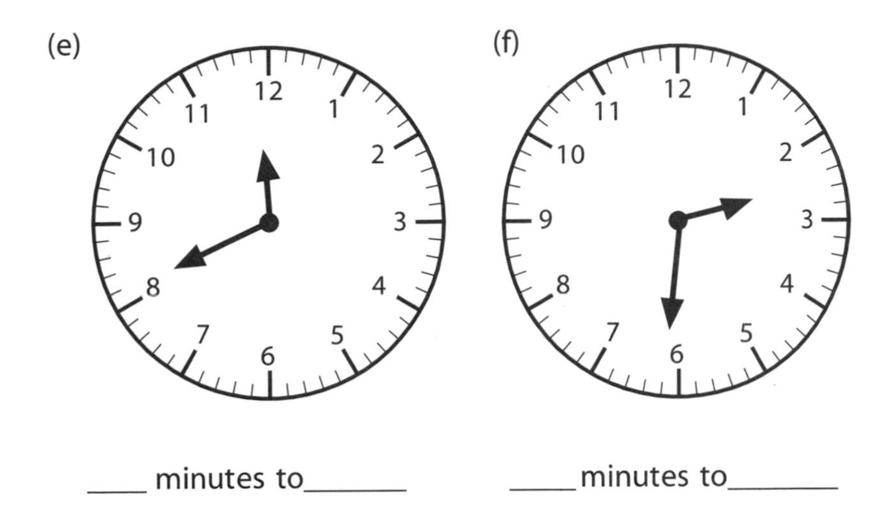

____ minutes to_______ ____minutes to_______

The final stage is to ensure that your child can write times purely as number, e.g., 3.45 a.m.

Practice in writing down times as numbers in a.m and p.m

Write down the time:

(a) morning

(b) afternoon

_____________________ _____________________

(c)
afternoon

(d)
morning

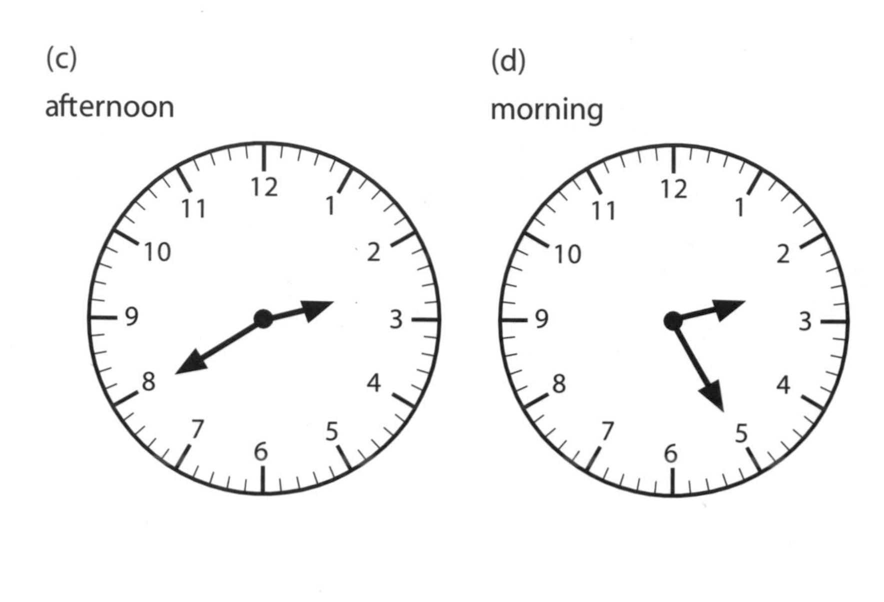

Draw the time on the clocks:

(a)

(b)

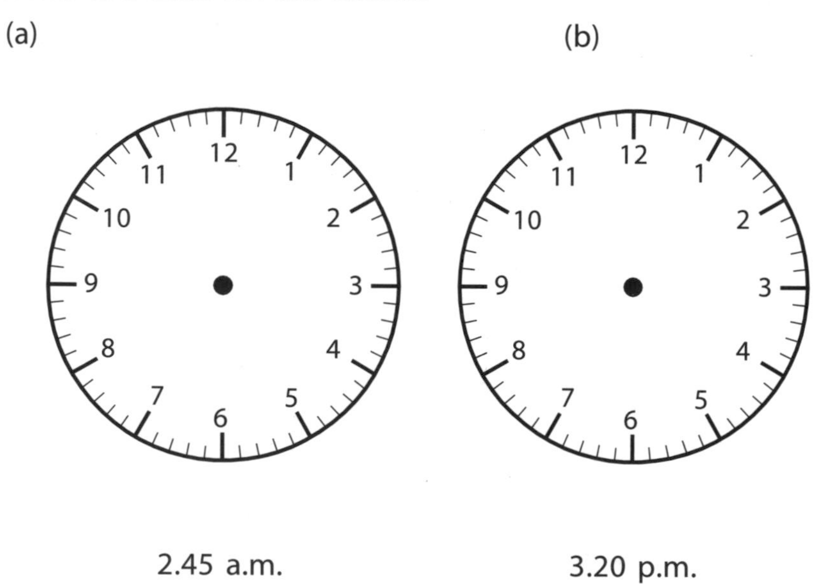

2.45 a.m.

3.20 p.m.

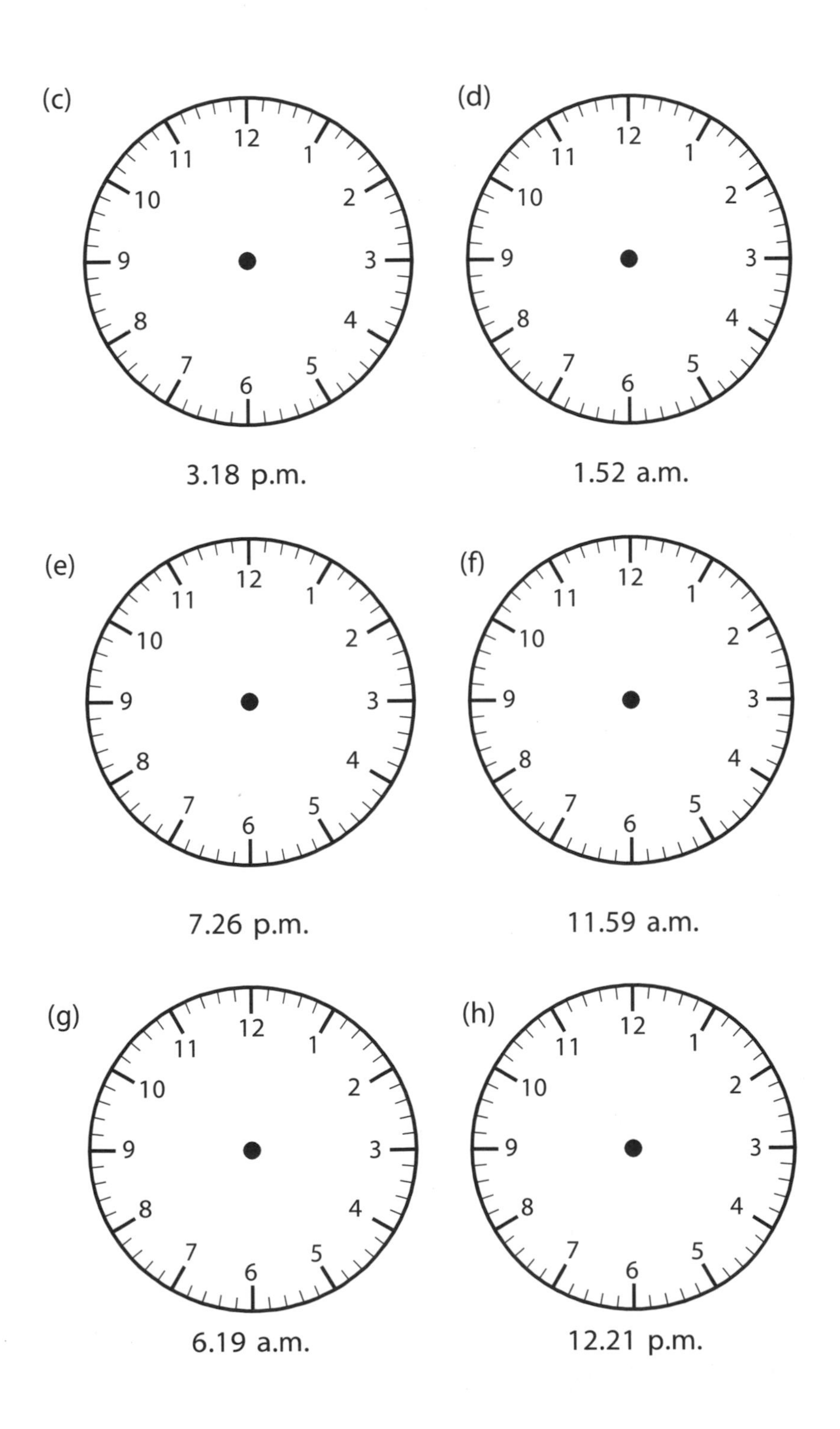
(c)
3.18 p.m.
(d)
1.52 a.m.
(e)
7.26 p.m.
(f)
11.59 a.m.
(g)
6.19 a.m.
(h)
12.21 p.m.

5 The 24-Hour Clock

For a child who advances quickly, the knowledge of the 24-hour clock is invaluable for everyday use, for example, for reading train timetables.

Learning the 24-hour clock is often pursued at an older age, but the idea of counting on from 12 mid-day, to 13 o'clock, 14 o'clock, etc, is easily understood by younger children.

RELATING TIMES OF THE DAY TO DAILY ACTIVITIES

At some point, your child will need to recognise that breakfast is at 8.00 a.m./9.00 a.m., lunch is at 12.30 p.m./1.00 p.m. and so on. One way to do this would be to set out **two** clocks, one representing the morning and the other the afternoon:

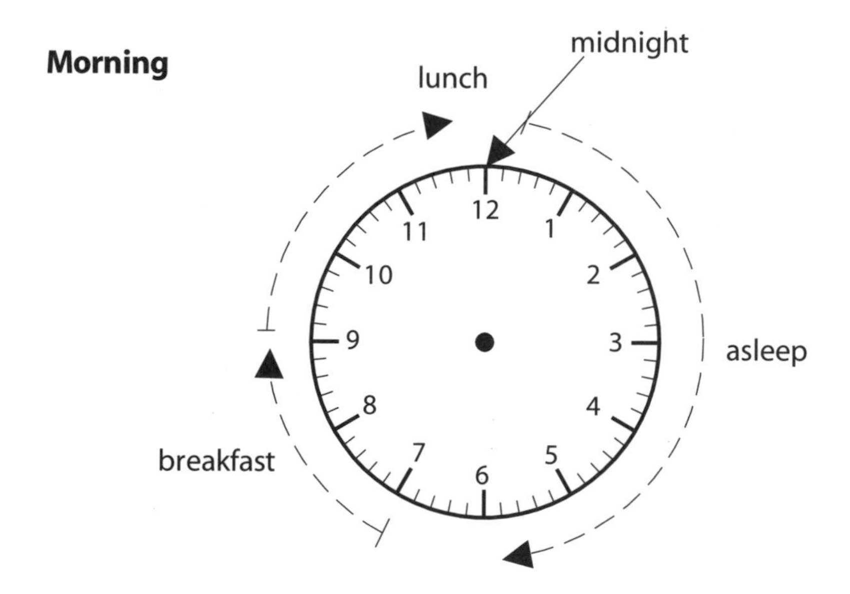

Afternoon

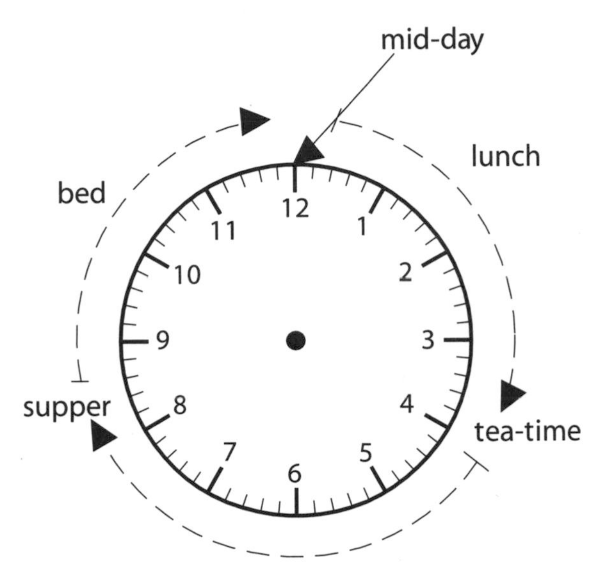

TRANSITION TO 24-HOUR TIMES

It helps in learning time if morning times are separate from afternoon times:

Morning

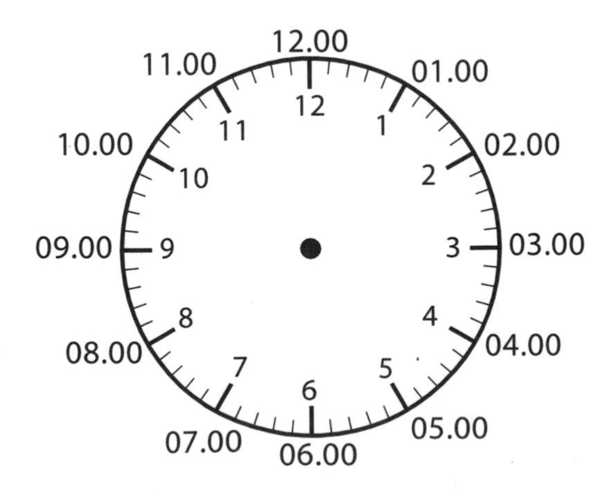

Afternoon

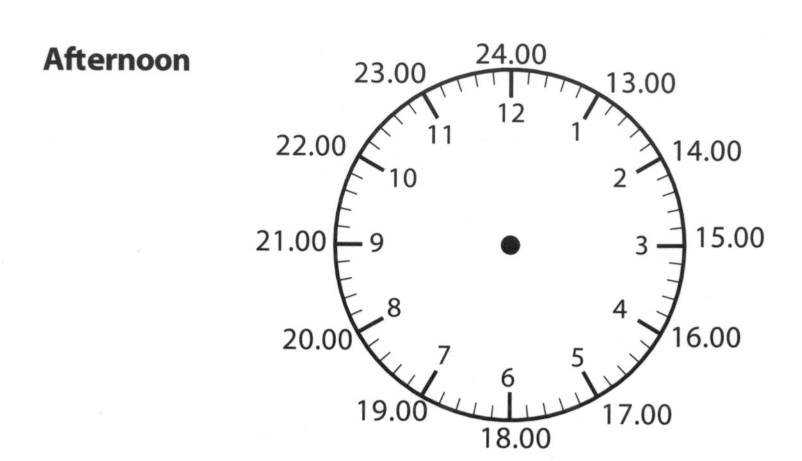

The 24-hour times can often be better understood and learnt, if in a 'list' pattern:

0.00 midnight
01.00
02.00
03.00
04.00
05.00
06.00
07.00 morning
08.00
09.00
10.00
11.00
12.00 mid-day
13.00
14.00
15.00

16.00	
17.00	evening
18.00	
19.00	
20.00	
21.00	
22.00	
23.00	
24.00	midnight

Practice at changing 24-hour times

For this age group, no attempt has been made to put in the minutes.

Change to 24-hour times

(a) 1.00 a.m. ______ (b) 8.00 a.m. ______

(c) 11.00 a.m. ______ (d) 12.00 a.m. ______

(e) 2.00 p.m. ______ (f) 6.00 p.m. ______

(g) 8.00 p.m. ______ (h) 9.00 p.m. ______

(i) 10 p.m. ______ (j) 11 p.m. ______

(k) 12 p.m.

____

Change to a.m. or p.m.

(a) 03.00

____

(b) 06.00

____

(c) 08.00

____

(d) 10.00

____

(e) 12.00

____

(f) 14.00

____

(g) 16.00

____

(h) 17.00

____

(i) 19.00

____

(j) 20.00

____

(k) 22.00

____

(l) 23.00

____

6 Finding How Long it is from One Time to Another

Recognising how long it is from one time to another is an important practical skill. How long is it before school starts? How long is it to tea-time? How long is that TV programme? How long is the maths lesson?

60 SECONDS IN ONE MINUTE

This is best illustrated with the aid of a stop-watch:

By noting how long it takes for the hand to make one circuit, your child will recognise 'how long' one minute is. You can show your child how to take their pulse, to see how many beats there are in one minute. Your child can also time how many seconds it takes to run down the garden and back.

60 MINUTES IN ONE HOUR

Minutes are marked on the clock, so this is a fairly easy exercise. Your child will need to appreciate and recognise the 'length' of one hour, half an hour, quarter of an hour, or three quarters of an hour, in particular.

It might be half an hour for a programme on TV, quarter of an hour to walk down to the shops or to school, or two hours to watch a film. It helps if you point out to your child the correspondence between the passing of time and the written number.

HOURS IN A DAY

How long? becomes of even greater significance when applied to hours of the day: How long is it to lunch? How long is it to break? How long to bed-time? How long before Grandma comes to visit?

Practice on time in the day

Get your child to count round on the clock.

How long is it from:

(a) 9.00 a.m. to 11.00 a.m? ________________ hours

(b) 6.00 a.m. to 10.00 a.m? ________________ hours

(c) 7.00 a.m. to 12.00 a.m? ________________ hours

(d) 2.00 p.m. to 5.00 p.m? ________________ hours

(e) 3.00 p.m. to 8.00 p.m? ________________ hours

(f) 6.00 p.m. to 11.00 p.m? ________________ hours

(g) 3.00 a.m. to 10.00 a.m? ________________ hours

(h) 4.00 a.m. to 12.00 a.m? ________________ hours

(i) 1.00 p.m. to 11.00 p.m? ________________ hours

(j) 11.00 a.m. to 1.00 p.m? ________________ hours

(k) 10.00 a.m. to 2.00 p.m? ________________ hours

(l) 9.00 a.m. to 1.00 p.m? ________________ hours

(m) 10.00 a.m. to 4.00 p.m? ________________ hours

(n) 9.00 a.m. to 5.00 p.m? ________________ hours

(o) 6.00 a.m. to 6.00 p.m? ________________ hours

(p) 7.00 a.m. to 9.00 p.m? ________________ hours

(q) 4.00 a.m. to 10.00 p.m? ________________ hours

COUNTING ON THE 24-HOUR CLOCK

How long from:

(a) 09.00 to 11.00? ________________ hours

(b) 06.00 to 10.00? ________________ hours

(c) 09.00 to 12.00? ________________ hours

(d) 13.00 to 17.00? ________________ hours

(e) 14.00 to 19.00? ________________ hours

(f) 15.00 to 23.00? ________________ hours

(g) 09.00 to 13.00? ________________ hours

(h) 10.00 to 16.00? ________________ hours

(i) 11.00 to 19.00? ________________ hours

(j) 10.00 to 21.00? ________________ hours

(k) 11.00 to 22.00? ________________ hours

(l) 09.00 to 23.00? ________________ hours

(m) 05.00 to 21.00? ________________ hours

HOURS AND MINUTES

This is more difficult for this age group, but for the more able child it provides a wealth of practice with number-adding, subtracting, and multiplying, as well as reinforcing knowledge of time.

Practice with hours and minutes

How long from:

Hours *Minutes*

(a) 8.00 a.m. to 8.15 a.m? ________________

(b) 7.00 a.m. to 7.45 a.m? ________________

(c) 1.00 p.m. to 1.26 p.m? ________________

(d) 1.00 p.m. to 1.53 p.m? ________________

(e) 8.15 a.m. to 8.35 a.m? ________________

(f) 8.30 a.m. to 8.46 a.m? ________________

(g) 10.45 a.m. to 10.56 a.m? ________________

(h) 2.23 p.m. to 2.51 p.m? ________________

(i) 2.40 p.m. to 3.10 p.m? ________________

(j) 3.45 p.m. to 4.23 p.m? ________________

(k) 1.20 p.m. to 3.29 p.m? ________________

(l) 2.23 p.m. to 5.38 p.m? ________________

(m) 2.15 p.m. to 6.09 p.m? ________________

(n) 11.14 a.m. to 1.26 p.m? ________________

(o) 10.20 a.m. to 3.40 p.m? ________________

(p) 9.34 a.m. to 4.19 p.m? ________________

(q) 8.23 a.m. to 5.31 p.m? ___________________

7 Practice for All Ages

By seven years of age it is reasonable for your child to tell the time to the minute, to write down that time in acceptable form, and to have some understanding of the 24-hour clock and time tables. Knowledge of time tables need not be extensive, and for some children can be left until they are older (e.g. nine years).

To try to 'push' knowledge at this stage will prove counterproductive, because the child will find the work irksome.

What follows in this chapter is further practice from recognising the o'clocks to practice with simple time tables.

Write the times beneath the clocks:

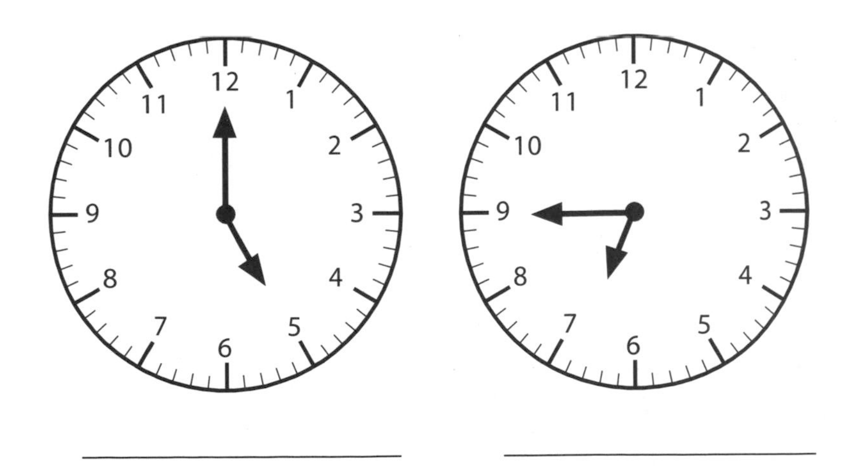

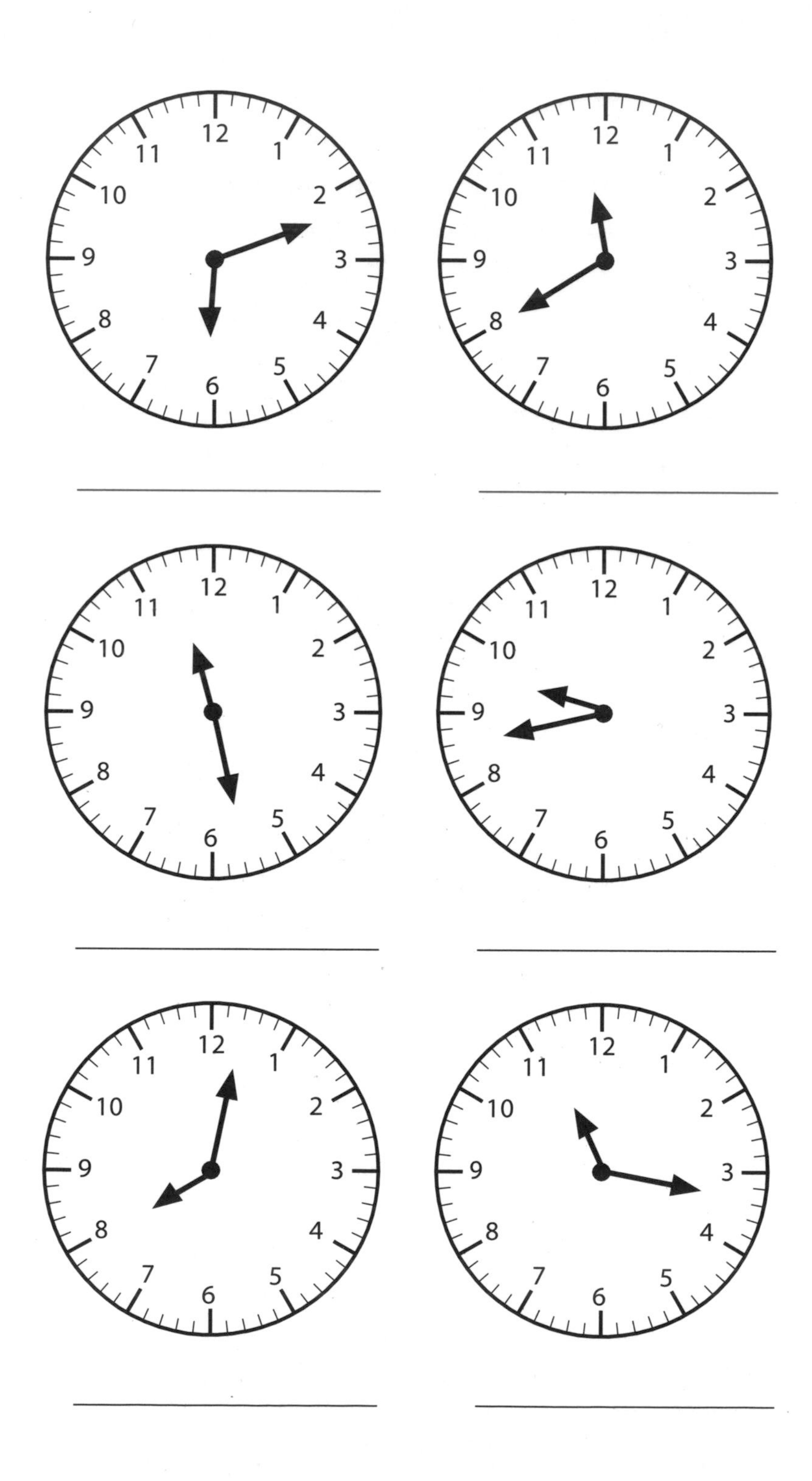
12
1
2
3
4
5
6
7
8
9
10
11

12
1
2
3
4
5
6
7
8
9
10
11

12
1
2
3
4
5
6
7
8
9
10
11

Put the hands on the clocks

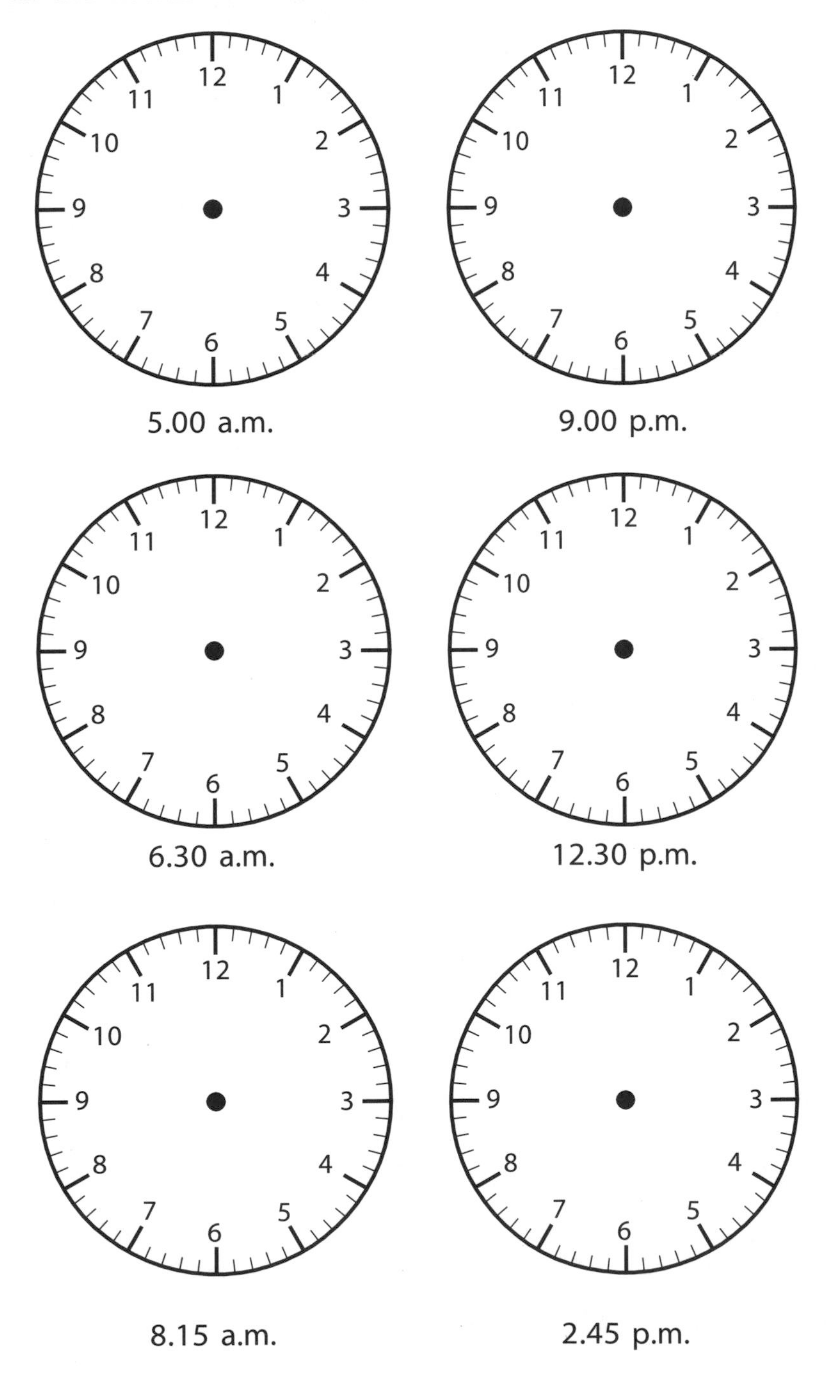

5.00 a.m.

9.00 p.m.

6.30 a.m.

12.30 p.m.

8.15 a.m.

2.45 p.m.

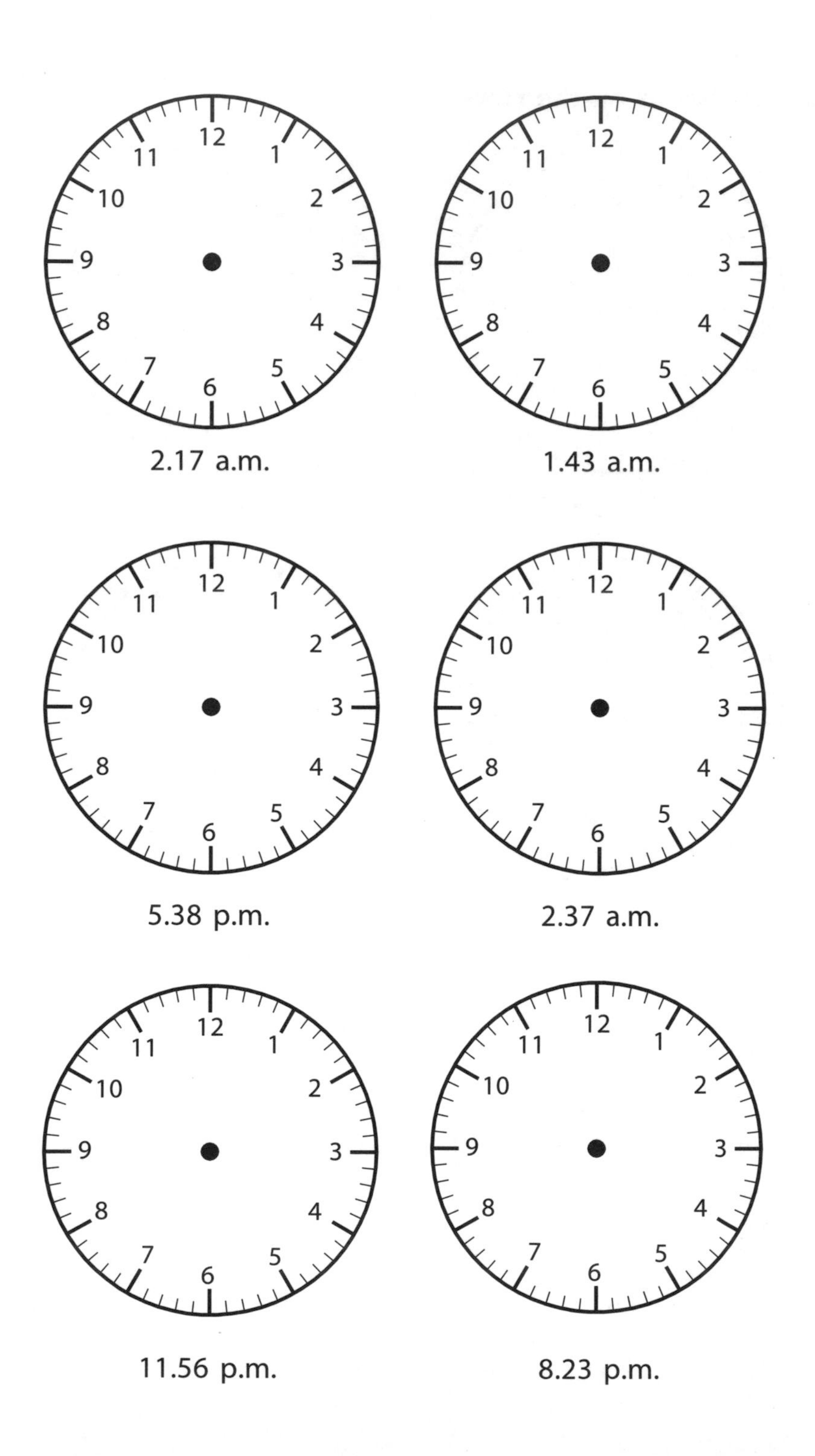

2.17 a.m.

1.43 a.m.

5.38 p.m.

2.37 a.m.

11.56 p.m.

8.23 p.m.

Change these times to 24-hour times

1	1.00 a.m.	____________	2	2.30 a.m.	____________
3	3.15 a.m.	____________	4	6.46 a.m.	____________
5	11.20 a.m.	____________	6	12.36 a.m.	____________
7	2.00 p.m.	____________	8	3.30 p.m.	____________
9	4.15 p.m.	____________	10	5.45 p.m.	____________
11	8.18 p.m.	____________	12	10.34 p.m.	____________

Change these times to a.m. or p.m.

1	01.00	____________	2	02.30	____________
3	04.45	____________	4	07.15	____________
5	11.14	____________	6	12.42	____________
7	14.28	____________	8	17.36	____________
9	21.19	____________	10	23.37	____________
11	04.37	____________	12	13.14	____________

This chapter and previous chapters have provided extensive practice in telling the time. The order of the practice examples at all times is sequential, a form of organising of learning that makes learning easy and fast.

The one thing that this book cannot do is to tell you how fast your child will learn. That part of learning must be left to your common sense and judgement. If you assess your child correctly, they will learn to the maximum of their ability, and will be happy and confident as they do so.

Extra Tests and Examples

Can you say which are o'clocks, half-pasts and quarters?

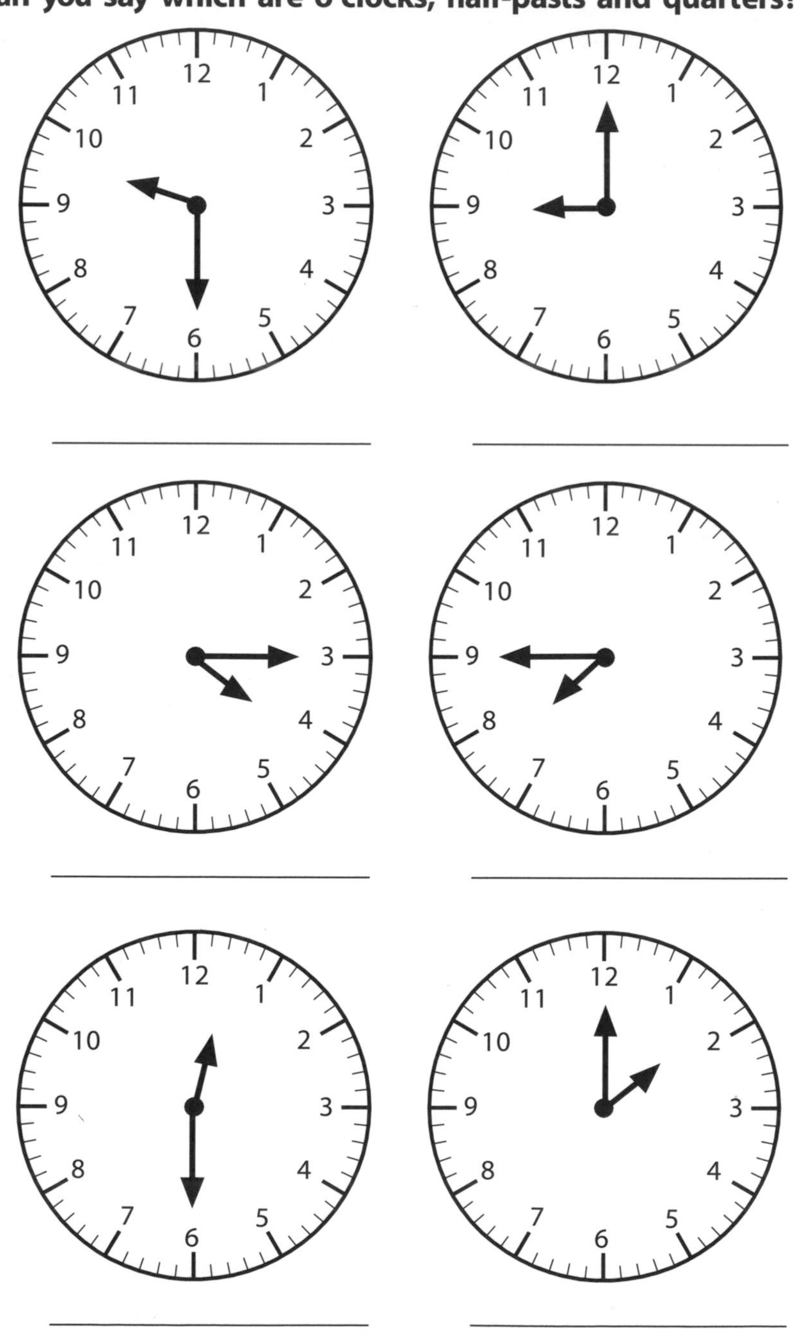

Counting round the clock

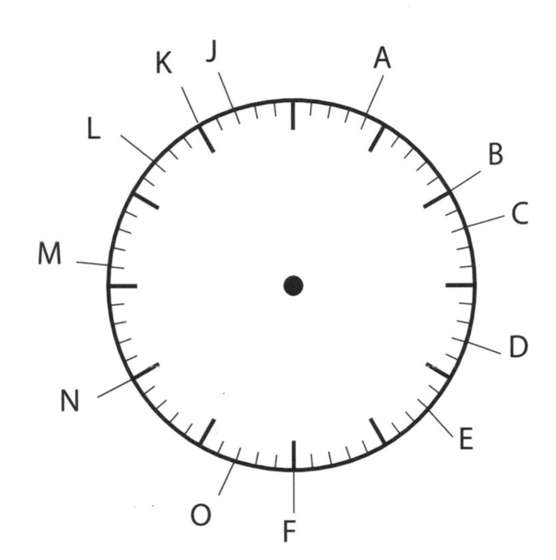

How far is it from the 12 point to:

(a) A ____________ (b) B________________ (c) C________________

(d) D ____________ (e) E________________ (f) F________________

How far is it back to:

(j) J ____________ (k) K________________ (l) L________________

(m) M ____________ (n) N________________ (o) O________________

Further practice of the 5 times table:

(a) 3 x 5 =________ (b) 6 x 5 =________ (c) 1 x 5 =________

(d) 10 x 5 =________ (e) 2 x 5 =________ (f) 9 x 5 =________

(g) 4 x 5 =________ (h) 7 x 5 =________ (i) 12 x 5 =________

(j) 5 x 5 =________ (k) 8 x 5 =________ (l) 11 x 5 =________

Minutes to and minutes past:

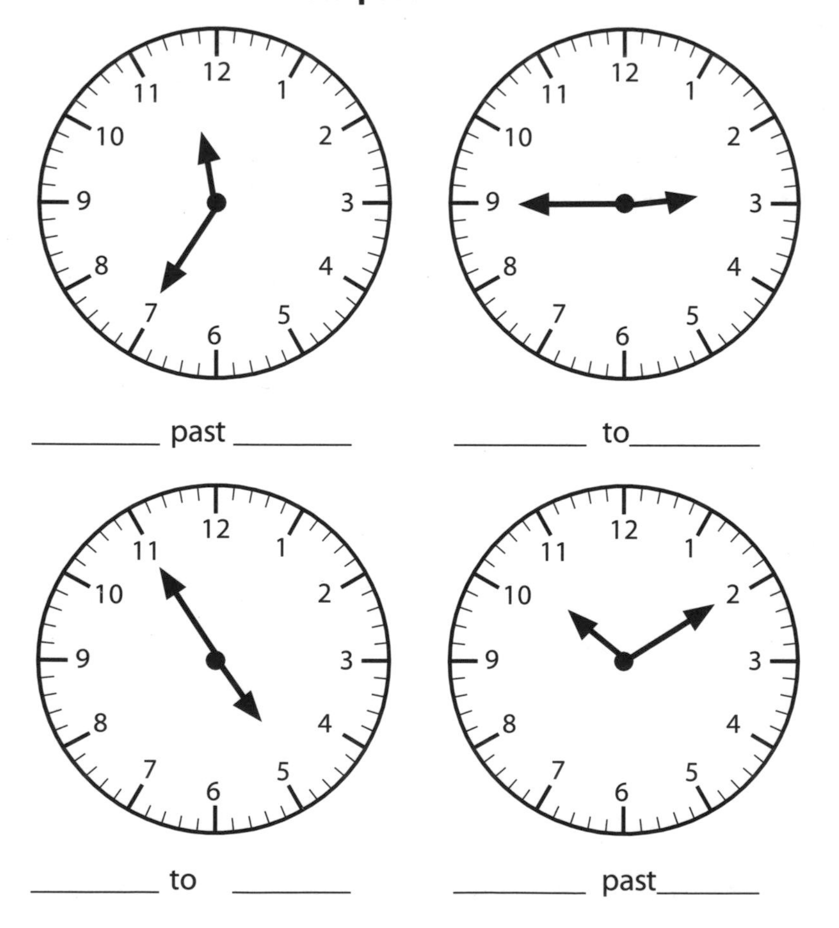

Practise writing down the time in a.m. and p.m:

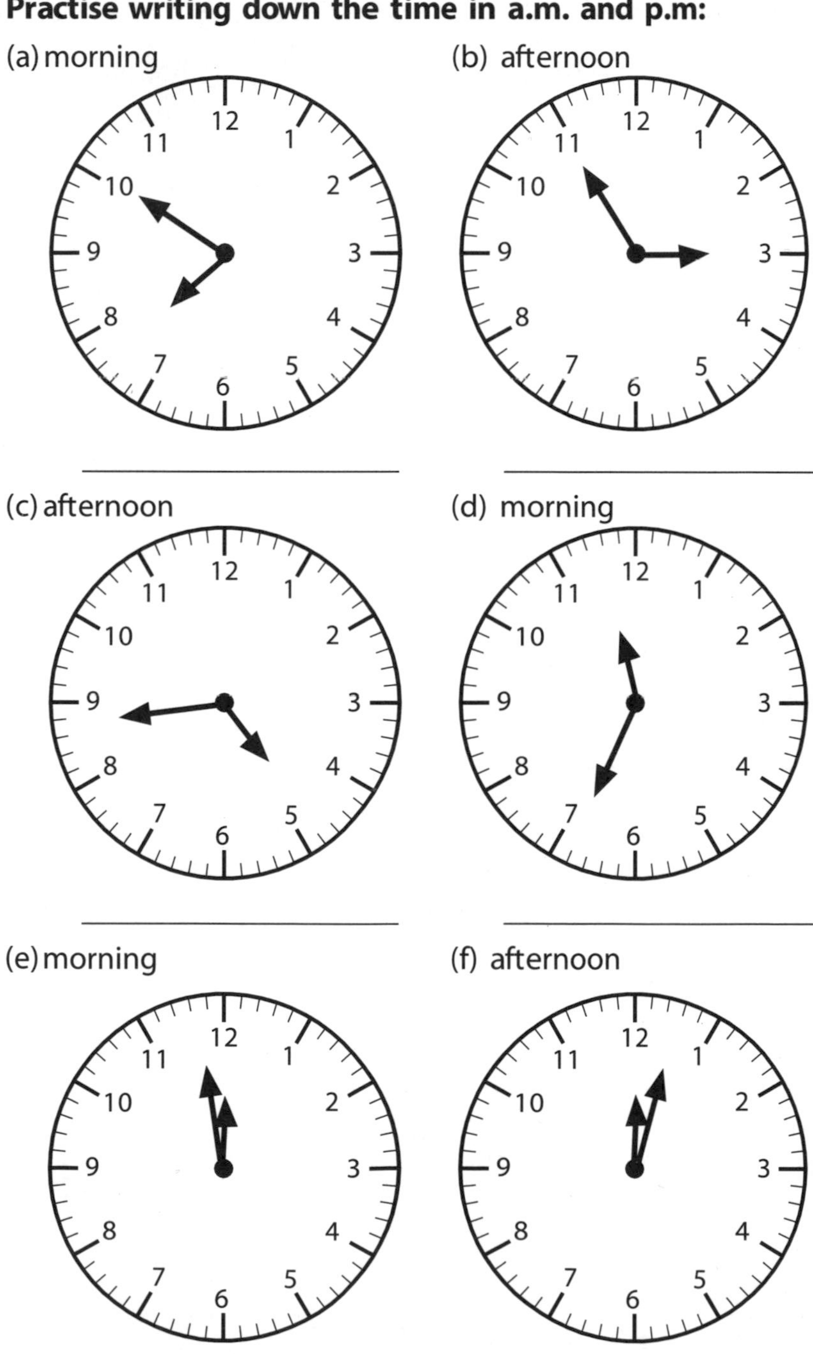

How long is it from:

(a) 2.00 p.m. to 5.00 p.m. ________________ hours

(b) 1.00 a.m. to 6.00 a.m. ________________ hours

(c) 6.00 p.m. to 10.00 p.m. ________________ hours

(d) 4.00 a.m. to 9.00 a.m. ________________ hours

Now in hours and minutes:

	Hours	Minutes
(a) 6.30 a.m. to 8.00 a.m.	________	________
(b) 2.20 p.m. to 4.30 p.m.	________	________
(c) 11.06 a.m. to 11.30 a.m.	________	________
(d) 8.25 p.m. to 9.15 p.m.	________	________

Change these to 24-hour times:

(a) 3.00 p.m. ________________

(b) 9.00 a.m. ________________

(c) 6.00 p.m. ________________

(d) 12.00 p.m. ________________

(e) 1.00 a.m. ________________

(f) 8.00 p.m. ________________

Change to a.m. or p.m:

(a) 19.00 ________________

(b) 02.00 ________________

(c) 16.00 ________________

(d) 08.00 ________________

(e) 21.00 ________________

(f) 23.00 ________________